How inappropriate to call this planet Earth when it is quite clearly Ocean.

— *Arthur C. Clarke*

OCEANOGRAPHERS SEARCH FOR A MYSTERIOUS CURRENT

TO THE DENMARK STRAIT

DALLAS MURPHY

Photography
Rachel Fletcher
Sindre Skrede

Video
Ben Harden

Chief Scientist
Robert Pickart

IN COLLABORATION WITH WOODS HOLE OCEANOGRAPHIC INSTITUTION

BURFORD BOOKS

This material is based upon work supported by the
National Science Foundation under award number: OCE-0959381.
Any opinions, findings, and conclusions or recommendations expressed
in this publication are those of the author(s) and do not necessarily reflect
the views of the National Science Foundation.

Burford Books
101 East State St., #301
Ithaca, NY 14850
www.burfordbooks.com

10 9 8 7 6 5 4 3 2 1

Library of Congress Cataloging-in-Publication Data
is on file with the Library of Congress.
ISBN: 978-1-58080-173-7

Jacket and text design by Melanie Roher, Roher/Sprague Partners
Printed and bound in China by C&C Offset Co., Ltd.

THANKS

It has been my great pleasure to have sailed aboard R/V *Knorr* on four expeditions in the Arctic and the Indian Ocean. The officers and crew have treated me like one of their own. I'll not forget their generosity. I owe thanks also to Eugenia Leftwich for her expert copyediting, to Sue Ann Sprague for dozens of tasks that benefited the book, and to Peter Burford, a true friend of the project. It has been my pleasure to work with Melanie Roher. Not only did she beautifully design *To the Denmark Strait*, she deftly produced it from start to finish. For that I offer my admiration as well as my thanks.

Finally, all my thanks and respect to Bob Pickart, who made it possible.

Dallas Murphy

CONTENTS

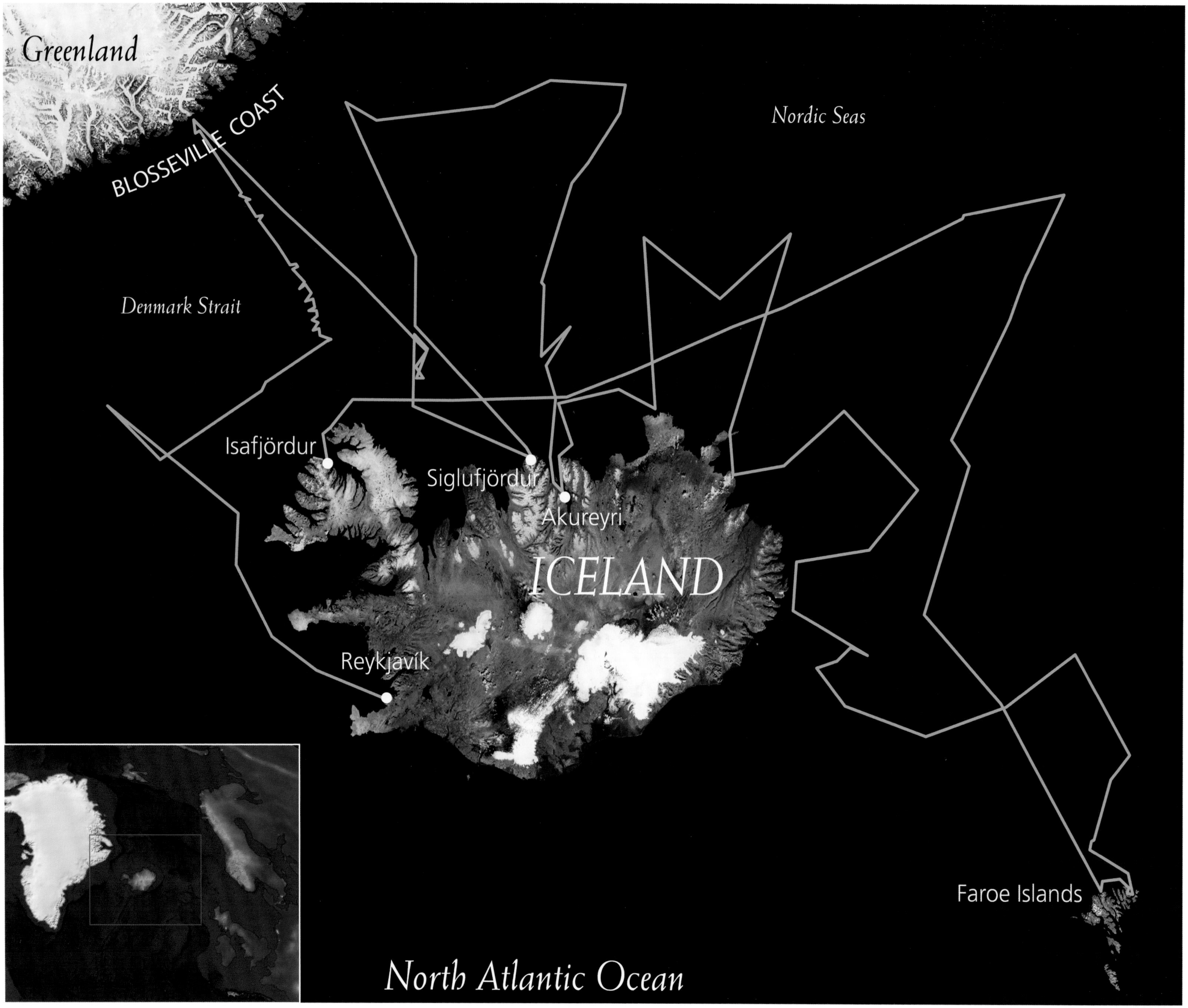

Greenland

BLOSSEVILLE COAST

Nordic Seas

Denmark Strait

Isafjördur

Siglufjördur

Akureyri

ICELAND

Reykjavík

Faroe Islands

North Atlantic Ocean

INTRODUCTION

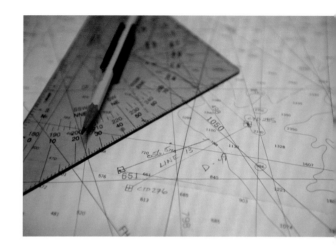

The ocean-research vessel *Knorr* put to sea from Reykjavik, Iceland, on a calm, sunny afternoon in August 2011. She returned to Iceland after thirty days and 3,812 nautical miles in Arctic waters to complete an extraordinary oceanographic expedition. This book is the story of that expedition. In its intimacy and its extent, *To the Denmark Strait* offers an unprecedented look at how ocean science is conducted and life is lived at sea aboard a world-class research vessel.

This world of at-sea oceanography is seldom witnessed by outsiders for the simple reason that it takes place far offshore aboard relatively small ships. Many people are unaware that dedicated scientists and their teams are out there right now in the 71 percent of Earth's surface covered by saltwater seeking to understand how the ocean works and what signals it's sending about the future state of our climate. In response to the brute difficulty of measuring oceans, specialized techniques, creative nautical adaptations, and particular social customs have evolved. People work on deck day and night operating heavy machinery, making fine-tolerance measurements, and deploying specialized instruments while dodging boarding seas long after conventional ships have battened down and dogged their weather doors.

Opposite: The track of R/V *Knorr* covering 3,812 nautical miles.

Top: Chief Scientist Bob Pickart on the bridge.
Bottom: The photographers off on a small-boat excursion.

Oceanography happens beyond the public's sight and ken, the very people who, with a small sliver of their taxes to the National Science Foundation, help pay for it. That's why Chief Scientist Bob Pickart encouraged the production of this book.

His Denmark Strait expedition was special on three counts. First, the science: All oceanographers put to sea with a specific purpose, say, to collect data about a particular region or current, and then they carry the data ashore for explication. This often takes years. In this case, however, Bob, from the Woods Hole Oceanographic Institution, and his colleagues went to sea with a specific hypothesis to test. If it held, the standing paradigm describing the climatically crucial circulation through the Denmark Strait separating Iceland from Greenland would need significant modification.

The hypothesis swung around the discovery in 1999 of a new current by Bob's Icelandic colleagues Hédinn Valdimarsson and Steingrímur Jónsson. That doesn't happen very often; the discovery of a new current draws attention, and people were skeptical. A 2008 expedition led by Pickart aboard this very ship proved that the current did in fact exist, and it wasn't just any new current. It flowed into the Denmark Strait, and, for reasons this book explains, that made it immediately significant. But where did it come from? What caused it? Pickart and his colleagues had a hypothesis explaining both. But the ocean, rough, vast, and variable, generally defies study; this one is downright bellicose. Big chunks of the schedule could be blown out. In 2008, we lost a week to hurricane-force winds. Also, equipment and ships under constant exercise and stress sometimes break. But if all held fast and the weather remained reasonable, if they actually managed to accomplish their measurements, Pickart and his international team would know whether their hypothesis was right or wrong by the close of this cruise. Inherent drama is not typical in physical oceanography.

Second, there was the "tourism." Members of most deep-sea expeditions see nothing of land except on departure and return. We, however, in the service of science, were treated to close-aboard views of some of the most magnificent high-latitude islands this side of Antarctica. There was East Greenland's fantastic, unexplored Blosseville Coast, where we

Bob with his co-principal investigators.
Left to right: Hédinn Valdimarsson, Robert Pickart, Laura de Steur, Kjetil Våge.

saw cathedral-sized icebergs, sometimes only as haunting specters in the fog, other times in dazzling sunlight, and we repeatedly launched small boats for close-up photography and video. We called twice on Iceland's mountainous, fjord-riven north coast. And then we visited the remote Faroe Islands, a place so unlikely in its isolation and so spectacularly exotic that we doubted the reality before our eyes.

Finally, we documented this expedition to our knowledge more extensively than any before it. The essays and images that comprise *To the Denmark Strait* were written and shot during the course of the cruise. We've tried to preserve that daily sense of excited immediacy as the best means of vicariously bringing the reader aboard. With three separate mediums—prose, still photography, video —*To the Denmark Strait* portrays the facts and feel of life aboard a ship dedicated to understanding how the ocean works, for its own sake and for its braided relationship to our climate.

Dallas Murphy
New York, NY 2012

UNDERWAY, AGAIN

With a promise not to dwell on it, I'll cop to a sometimes sentimental romanticism about seagoing ships and boats. So suffice it to say it was a pleasure to join this fine old ship *Knorr* and her people at the wharf in Reykjavik, Iceland. I hadn't been aboard since 2010 in the Indian Ocean and before that, in 2008, in these very waters. On that cruise, Bob Pickart and Kjetil Våge confirmed the existence of the North Icelandic Jet a few years after its initial discovery by Hédinn Valdimarsson and Steingrímur Jónsson. On this cruise, they mean to find its origin and cause.

As I stepped off the gangway onto the main deck, my friend Pete, chief engineer, said," Let me show you something."

"What?"

"Don't ask, just come."

I dropped my gear against a bulkhead and followed him up a ladder to the 01 deck.

"Remember this?"

"That's not—"

"Yes. It is."

In addition to the search for the North Icelandic Jet, the objective of this cruise is to lay moorings across the Denmark Strait between Iceland and Greenland. The foundation of an oceanographic mooring is a stout wire with an anchor at the bottom and, near the surface, a top float 42 inches in diameter delivering some 2,000 pounds of buoyancy to keep the wire vertical. Scientists mount arrays of instruments on the wire to measure water temperature, salinity, and current velocity, among other things. About ten spanking-new top floats still in their protective plastic wrap were mounted on the 01

Bob Pickart and Kjetil Våge confirmed the existence of the North Icelandic Jet a few years after its initial discovery by Hédinn Valdimarsson and Steingrímur Jónsson. On this cruise, they mean to find its origin and cause.

Opposite: R/V *Knorr* in Reykjavik Harbor, ready to go.

The errant mooring ball, washed overboard on the 2008 cruise.

deck, but one, separate from the rest, was badly battered and heavily encrusted with barnacles and sundry marine growth.

"I bet you never expected to see that again," said Pete.

The weather on that 2008 cruise was spectacularly violent, one deep storm after another marching out of the west as closely packed as beads on a necklace. During a particularly savage 70-knot blow, a boarding sea ripped this mooring ball from its deck mount. The heavy lashings didn't part; the wave tore away the metal mount welded to the side deck. The 2,000-pound ball had bounded aft along the starboard-side deck and stove in the steel bulwarks on the stern quarter before it went overboard. We're still telling the story of that storm. No one expected to see that expensive piece of gear ever again. But here it was.

"How? What—?"

"An Icelandic farmer found it on his land," said Pete.

There was pleasing congruity in this both personally and, more important, scientifically, one circle closing, another beginning. In a couple of hours, we'll be underway again heading back to those same waters where we were so harshly treated three years ago. But then, that was in October. Now, in August, the storm track should be less active in our study area, but of course *should* and *will* are two different concepts when ocean and wind are involved, especially above the Arctic Circle.

. . . .

It's 1015, August 22, and we've just cleared the inner breakwater in a fine drizzle, gunmetal-gray clouds hiding the snowy mountaintops. Captain Kent Sheasley, Second Mate Jen Hickey, and helmsman Jose Andrate are on the bridge with the Reykjavik Harbor pilot.

"Steer zero-five-zero," said the pilot.

"Zero-five-zero," Jose repeated according to age-old protocol.

"Keep that yellow buoy to port," said the pilot. In practical terms, this process was unnecessary, since *Knorr* has been in and out of Reykjavik many times, but regulations require.

The pilot boat approached, turned sharply, and took up station off our starboard side. Then, having passed the sea buoy, the pilot's job was done. He and the captain exchanged official paperwork, shook hands, and the pilot left the bridge.

"You can put her on the waypoint to the first station," said Jen. "Three-three-zero."

"Three-three-zero."

"The pilot's away," reported a seaman by radio from the main deck.

Then Jen stepped out onto the bridge wing and struck the "pilot aboard" flag from the mast. Let that act signal the official start of our cruise. It was 1040. The rain stopped.

A safety drill is scheduled for 1230, when we're to report to our muster stations with life jackets and survival suits. The wind is very light, and *Knorr* begins to roll with it ever so slightly. It feels good. We're at sea again.

Knorr working off the coast of Iceland.

A lone gannet as escort.

In a couple of hours, we'll be underway again, heading back to those same waters where we were so harshly treated three years ago.

THE IDENTITY OF WATER

It's blowing only 25 knots this morning from north-northeast, but the rumpled sea state suggests heavier wind. That's probably because there's no land away in that direction, only ocean. The distance over which wind blows unimpeded by land is called *fetch*. Long fetch results in higher waves for the same wind velocity as shorter fetch. I was just up on the bridge listening to the nautical talk—I've missed it—and thinking about the difference between the mariner's view of the ocean and the oceanographer's.

Knorr's bridge perches 45 feet above the waterline. From that height of eye, says the *Nautical Almanac*, we can see a circle of water with a 7.8-mile radius before Earth curves away. The bridge watch pays very close attention to that water, particularly the quadrant ahead. The average depth of the Atlantic is about 12,000 feet, but at sea, where there's no bottom to hit, mariners need only consider the surface of the sea to do their jobs successfully. To do his, however, the oceanographer needs to understand all of it. And like all Earth scientists, oceanographers understand their subject first by measuring it. But before we address the question of just how you measure an ocean, we should address a couple of fundamental ocean principles in order to appreciate the trajectory and the importance of this voyage:

First, the ocean is alive with constant motion. Wave motion is readily observable even from the beach, by those Herman Melville termed "water gazers." Tidal movement is also visible, and a kid learns of it when high tide encroaches and dissolves his sand castle. But that barely scratches the surface of ocean dynamics. Dozens of currents, inaccessible to the naked eye, course through the body of the world ocean. Some flow on the surface driven largely by global wind belts such as the tropical trade winds and the prevailing westerlies in the mid-latitudes. Some surface currents are

To apprehend the true wonder of the ocean-current system, we need to imagine the ocean alive with motion, currents flowing on the surface and through the deep darks, never ceasing, transporting heat and surface water from the tropics to the Arctic and returning cold water at depth.

Opposite: The Denmark Strait in a less-welcoming mood.

In this body of water between Greenland and Norway occurs one of the most complex and important arcs of circulation in the entire world ocean. The stability of global climate depends on the integrity of the circulation in the Nordic Seas.

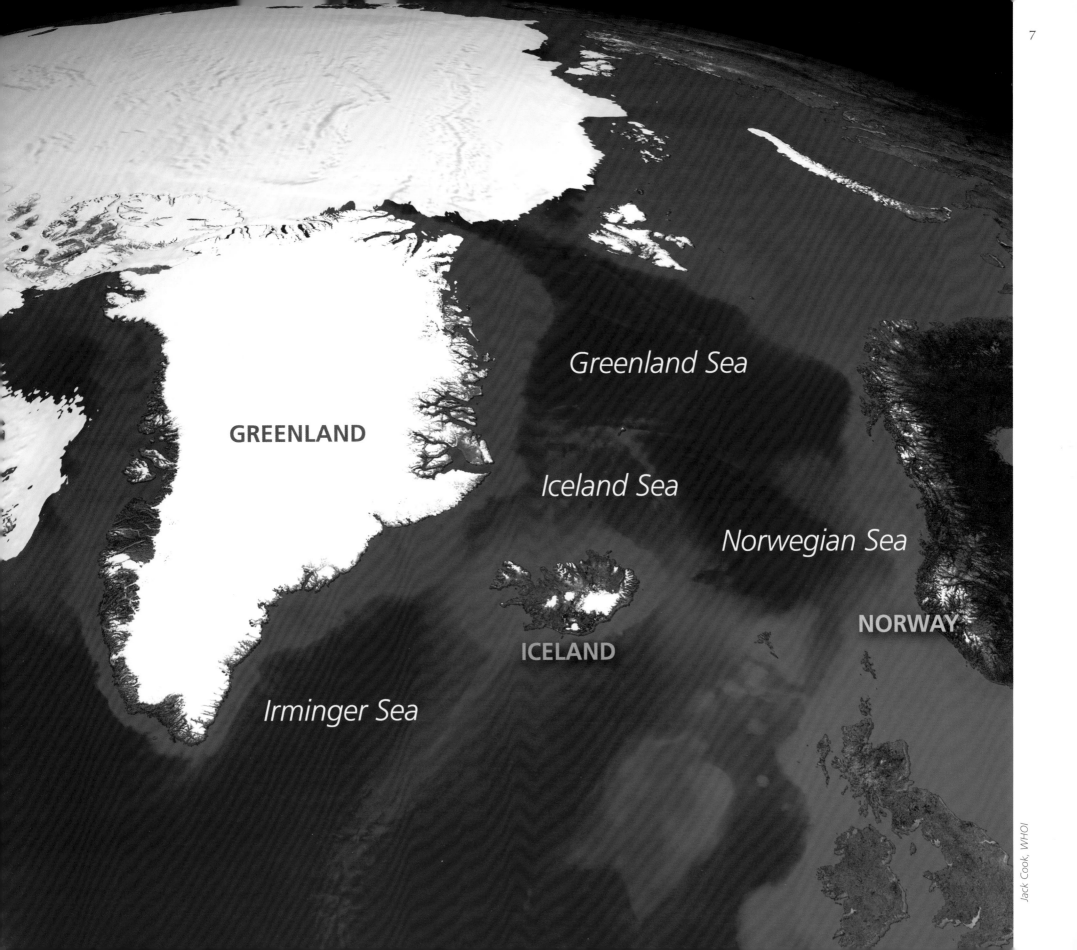

GREENLAND

Greenland Sea

Iceland Sea

Norwegian Sea

NORWAY

ICELAND

Irminger Sea

cold, some warm. Some flow fast, transporting lots of water in a relatively thin stream, while others amble languidly over broad swaths of ocean. Some currents flow through the deep ocean, in the intermediate depths, and along the bottom several miles beneath the waves. Together, they're called *ocean currents,* a technical term referring to a particular type of water moving permanently in a single direction. Ocean currents will wobble in velocity and wander in direction, but in the mean they always flow in one direction. We'll talk much more as the trip evolves about the climatic significance of ocean currents. But for now it's important to recognize that by transporting warm water from the tropics to the Arctic and cold water from the Arctic back to the tropics, ocean currents moderate climate, thus rendering habitable broad zones of the Earth that would otherwise be too hot or too cold to support present ecologies, humans included.

Second, the ocean is not a big pool of undifferentiated water. In most ways the ocean, rough and vast in three dimensions, defies our efforts to understand it. But in one generous respect it cooperates with the oceanographer's objectives: Water masses—currents—retain their individual fingerprints as they course around the world. Oceanographers can distinguish one water mass from its neighbor or from half a world away by measuring its temperature, salinity, and velocity. That makes the job sound easy when it's not. It's difficult, arduous, and occasionally dangerous; plus, you get seasick. But it can be done. And on this cruise we'll be doing it in places where it's not been done before.

The bridge, the "brains" of the ship.

Kjetil Våge at work in the main lab.

In most ways the ocean, rough and vast in three dimensions, defies our efforts to understand it.

Let's leave it there for now and talk later about the devices with which ocean scientists measure their subject. Well, wait, let's add one more aspect related not so directly to the science, but to our appreciation of it. Oceanography requires imagination. Geologists, botanists, and biologists can see their subject, if sometimes with the aid of magnification. But not the oceanographer. We look out from *Knorr's* bridge and see—water, and not very much of it at that. Nautically, that's enough; oceanographically, it's nothing. Our eyes are useless, oceanographically speaking. To apprehend the true wonder of the ocean-current system, we need to imagine the ocean alive with motion, currents flowing on the surface and through the deep darks, never ceasing, transporting heat and surface water from the tropics to the Arctic and returning cold water at depth. We're talking here only about the North Atlantic and the adjoining Nordic Seas, but a variation on the theme is playing in all the world oceans every instant of every day.

Our first of many fog banks.

HOW TO MEASURE AN OCEAN, PART ONE

We're just a tick north of 66 degrees North latitude—that's 33 miles south of the Arctic Circle, at the northern end of the Denmark Strait—and the rare sun just came out. Seas are calm by local standards, and this makes the mooring technicians happy. Mooring operations are just getting started. The techs have laid out their tools, hardware, current meters, and measuring devices on the flat aft deck; they've spooled the specialized oceanographic wire onto the big winch drum and positioned the synaptic-foam top float, the first piece of the mooring structure to go overboard, on the brink of the transom. That *Knorr* is barely making way means we're very close now to the position Bob has chosen for his first measuring point. The techs move with practiced precision about the deck—between them, they've planted moorings in all the world's oceans—but everyone is taking very seriously this difficult, potentially dangerous process.

It's called a mooring because, like all kinds of moorings, it's anchored to the bottom, but the word denotes its structure, not its purpose. The structure consists of a wire strung between a two-ton anchor and one of those flotation balls set near the surface to keep the wire vertical in the water column. Onto this wire, scientists can string a top-to-bottom array of measuring devices that automatically record data for a year and more. That's the major advantage of the mooring—it can remain in the water for long periods, especially important since time resolves the constant, small-scale variations inherent to ocean-current behavior, leaving a mean. So the question is, what of the ocean do scientists want to measure with these things, and how?

Yesterday I mentioned that currents, as they flow around the globe, retain an identifiable fingerprint in the form of temperature and salinity (T/S). If scientists can identify the water contained in a current, they can follow its path and determine its behavior. To do so, Bob and his colleagues

Preparing to deploy the first mooring.

Opposite: Knorr looking aft from the bow.

Tools of oceanography.
Top: The acoustic releases. *Middle:* A current meter.
Bottom: Acoustic Doppler Current Profilers.

attach to the wire miniaturized devices called CTDs (about which more later) to measure T/S profiles and store the information on data cards, rather like those in digital cameras, for later retrieval. (These small CTDs are referred to by the pros as Sea-Birds, after their manufacturer.)

In addition to its T/S profile, scientists want to know how fast the water is moving. From this knowledge, they can calculate the current's volumetric transport over the course of one year, and thus determine the quantity of heat being transported, which speaks to that relationship between the ocean and climate. Small current meters are attached to the wire at various depths to register the velocity of the passing water. This is a simple and direct means to address the objective, but in itself not an entirely complete solution. You see, because that top float cannot remain on the surface, where it would be jeopardized by passing fishing vessels and ice, the wire length is calculated so as to leave the top float some 100 meters beneath the surface. However, that top 100 meters is important and must be measured. That's where the Acoustic Doppler Current Profiler (ADCP) comes into its own.

You've probably noticed how the audible pitch of a fire-engine siren or locomotive changes as it approaches and passes, the so-called Doppler shift. From its mount on the top float, the "upward-looking" ADCP shoots pulses of sound waves into the water which, when they encounter particles such as sediment and plankton drifting with the current, bounce back at a slightly but measurably different frequency. From the difference, you can calculate the velocity of flow. Additionally, ADCPs are built into *Knorr's* hull, constantly pinging the water throughout the cruise, and when you factor out the ship's speed and heading, you have current velocity. However, because the currents vary widely in velocity over time, the shipboard ADCPs cannot substitute for the lowered ADCPs mounted on moorings. Also, shipboard ADCPs are limited in range to about 400 meters.

But wait, you might be asking, if the top float to which the entire mooring structure is attached remains 100 meters beneath the surface, how do you retrieve it? Many oceanographic measuring devices, which have been adapted from Cold War anti-submarine

Mooring top floats waiting to go overboard.

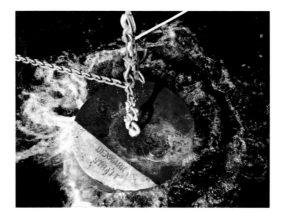

Deck work begins for the first mooring deployment.

Opposite, bottom: Positioning the anchor for deployment.
Right: Attaching the "hardhats" for additional floatation.

"Hardhats" and acoustic releases going over.

*There's a saying in the business that when you put instruments in the water,
you no longer own them; the ocean does, and it may or may not give them back.*

technology, depend on sonar, because sound waves travel quite effectively through seawater. The ADCP is an obvious example. The "acoustic release" is another. This clever device is attached to the wire at the bottom just above the anchor. When the time comes to retrieve the mooring to collect its stored data and change batteries, a technician "talks" to the release in an acoustic code from a portable computer called the deck box. The release responds in code, saying, essentially, I'm awake and ready for instructions. The next code tells it to let go. The jaws that connect it to the anchor open, and the mooring—minus the anchor—floats to the surface to be retrieved and brought aboard.

The mooring team, led by John Kemp and Jim Ryder, two of the best in the business, have lowered the top float into the water, and now they're paying out the wire as the ship steams slowly forward. The techs are stopping the wire to attach instruments at various predetermined points. Now the yellow float streams so far behind the ship it's visible only with binoculars. Soon they'll attach the acoustic release and splash the anchor on mooring number one. When the last one goes over the stern, they will have laid for the first time ever a string of moorings capable of measuring the flow through the entire Denmark Strait—if, that is, all goes according to plan. It doesn't always.

There's a saying in the business that when you put instruments in the water, you no longer own them; the ocean does, and it may or may not give them back. Or expressed differently, "If you can't afford to lose it, don't put it over the side." We were asked by German oceanographers to try to retrieve a mooring they had deployed in this area a year or so ago. Their acoustic release had responded to the wake-up call but didn't let go. They had tried dragging for it with grappling hooks, but to no avail, and had to leave the mooring behind. So they supplied the release code and asked us to give it a last try. No one aboard here gave it much hope, but we dutifully steamed to the position and talked to the release. Sure enough, it responded but still refused to relinquish its hold. That mooring is gone, probably forever, many thousands of euros lost, not to mention the still-more-valuable data it had been collecting and will continue to collect until the batteries expire. But that's how it goes. Everyone hates to lose stuff, but everyone expects to.

The top float with an ADCP mounted on it going over the side.

IMAGINING OCEANS

To appreciate the elegant beauty of nature's great ocean-atmosphere systems, we need to cast our minds out over vast distances and into opaque depths. And to begin to grasp the vital role played by the Denmark Strait in the circulatory system, let's take an imaginary—but quite naturalistic—drift from the subtropics back up here to the subarctic aboard a small boat (we'll assume fair weather), no power, letting the current do the work. The Gulf Stream in the Straits of Florida might be a good place to start for the sake of context and a sense of the interconnectedness of the ocean system.

The bruise-blue Gulf Stream water is blowing through the close quarters between Florida and the Bahamas at a steady four knots, transporting some 25 million cubic meters of water *every second* past Miami, Ft. Lauderdale, and Palm Beach. North of the Straits, we slow a bit, but the Gulf Stream, hugging the continental shelf, still sprints northward toward the bulge of Cape Hatteras, North Carolina, at high speed. If the weather's clear we'll get a glimpse of the sand dunes on Hatteras before the Stream carries us northeastward out into open ocean. It's still not definitive why the Gulf Stream departs the coast and heads toward Europe. Though it's called a surface current, the Gulf Stream's influence reaches down thousands of meters, so it's probably being steered northeastward by bottom topography, *bathymetry* in technical parlance. Now, unconstrained by land, in very deep water our current begins to meander freely, sometimes sinuously, but if we don't get spun off into one of the big eddies on either side of the main flow, we'll ride on toward the Grand Banks of Newfoundland and Europe beyond.

However, contrary to popular belief, the Gulf Stream proper never reaches Europe. The fast, clearly defined current begins to dissolve into filaments of flow that bend southward in a broad,

Contrary to popular belief, the Gulf Stream proper never reaches Europe.

Opposite: A fair-weather welcome to the Denmark Strait.

Ocean circulation is one of the great geophysical forces that shape the conditions of our world—that's why scientists study its dynamics.

Cold deep current

Warm surface current

slow drift bound back for the tropics. But we're not going that way. We've been entrained in an extension of the warm Gulf Stream called the North Atlantic Current (NAC) now veering off to the north from near the Grand Banks toward—and past—the British Isles. This is the warmth that, in collaboration with the west winds, nurtures those famous, incongruous palm trees in the south of England and in Scotland. They offer a perfect example of the relationship between ocean circulation and climate when we consider that at the same latitude as London, on the west side of the Atlantic, we find frozen Labrador. But as we pass north of Scotland, things are changing deep beneath the surface. Though our surface current hardly notices, we're passing over a submarine mountain range running from northern Scotland west, through Iceland, and on across the Denmark Strait to Greenland. In fact, it's called the

Greenland-Scotland Ridge. Crossing the ridge, we enter the Nordic Seas. Here the NAC gains a name change, to the Norwegian Atlantic Current, but that's just nomenclature. It's the same current, and its warmth graces Norway with ice-free conditions year-round north of the Arctic Circle.

Now, with the arrival of autumn, the current is quickly relinquishing its warmth to the cold air. Cold water is heavier—denser—than warm, and the tropical-origin salinity increases its density. Soon, with the onset of winter, the cold, salty water begins to sink. The trip's getting uncomfortable and dangerous now; it's time to go ashore and let the ocean circulate without us. Besides, the surface circulation is growing complex and confusing. There's no telling where our boat would go if we stayed aboard. It might drift off over the top of Norway into the Barents Sea. It might travel

Icebergs along the east coast of Greenland.

with an arm of the Norwegian Current over Spitsbergen into the High Arctic. Or perhaps it would flow around the edge of the basin—currents like to hug the edges of ocean basins—and down the East Greenland coast. In any case, an "ocean problem" has arisen.

That gargantuan quantity of Atlantic-origin water flowing into the Nordic Seas basin must now find a way out. A law of nature insists that if a quantity of water flows north, then an equal quantity must flow back south. If the water just piled up, Western Europe would have been submerged eons ago. But where can it go? Most of the southbound flow spills over the shallow, 600-meter sill in the Denmark Strait into the deep Atlantic. Then, through an extremely complex, only partially understood but magnificent process, this water forms up into a thin stream, the so-called Deep Western Boundary Current, which flows southward near the seabed and proceeds beneath the Gulf Stream toward the equator. Thus the circle is conserved.

The elegant system I've so cursorily described—the northward surface flow of warm, salty water from the Gulf Stream System and the return of cold water at depth—is called the Meridional Overturning Circulation (MOC). Should any segment of the circle be severed—say, by an influx of freshwater from the melting Greenland Ice Sheet or some other feedback from global warming—then climatic disaster will result. No serious scientist is going so far as to say that will happen, only that it *could* happen; the potential is inherent to the system. Ocean circulation is one of the great geophysical forces that shape the conditions of our world—that's reason enough for scientists to study its dynamics and, since we live in the world, reason enough for us to look over their shoulders as they do so. Besides, no one knows where the endeavor may lead. In the ocean may lie answers to questions humankind has yet not thought to ask.

PLAY DAY (FOR SOME)

"Want to go for a boat ride?" asked Captain Sheasley.

"Sure."

"Then grab a Mustang suit and come on."

The deck crew, with the sure-handed seamanship typical to this vessel, craned the small boat (a rigid-hull inflatable, or RHIB) off the 02 deck, lowered it to the main deck and secured it to the starboard rail for ease of boarding. The purpose of this trip was photography. Rob McGregor is shooting footage for a multipart *National Geographic* special about the ocean, and Sindre Skrede is shooting for a couple of Norwegian websites. They were not playing. Captain Sheasley, "the Skip," would drive the boat; he was sort of playing. Along for the ride, utterly superfluous, I was playing.

It's always interesting to observe the direct relationship between the weather and the general mood aboard ships and boats at sea. The weather yesterday was exquisite, high-latitude perfection. The wind was slack, the sea dead flat except for long, gently undulating swells out of the east. The sky was blue and cloudless except for a nonthreatening, half-formed front away to the west just dark enough to offer pleasing contrast. It looked like the gentle tropics if you ignored the jackets and wool hats and the Greenland iceberg floating on the horizon. (The Skip tells me that the new crew, having heard repeatedly the stories of warlike violence from the 2008 trip, is beginning to doubt our veracity.)

Jim and John, the hotshot WHOI mooring team, Kyle the bosun, with co-Principal Investigator (PI) Kjetil Våge, were assembling and deploying another mooring with their usual skilled choreography as we climbed aboard the RHIB. The crane operator lowered us into the water. Rob was geared up in his dry suit and fins, planning to shoot the mooring ops from in the water (temperature, 1.7° C, 35° F). Not a common event on research ships, this attracted attention from a bevy of scientists, off-duty techs, and crew lining the transom peering through camera lenses. It bespoke tourism.

Opposite and above: In the workboat to shoot a mooring launch from water level.

The water, which looks flat black from aboard ship, was gin clear—alluring, if only it weren't a tick or two above freezing.

The top float armed with an ADCP.

"Everybody hanging on?" asked the Skip. We were. He put the hammer down and we sped away astern toward the mooring ball bobbing in the swells.

By the typical deployment process, the big top float, loaded with an Acoustic Doppler Profiler ADCP and miniature CTD to measure temperature and salinity, goes overboard first, and as the techs add the appropriate instruments one by one to the wire, they pay out the ever-lengthening mooring. The ship steams ahead at about one knot to keep the ball and the wire streaming straight aft, and the procedure continues until all the instruments have been attached. Finally, the two-ton anchor is splashed overboard. Rob wanted shots of the ball in the water and especially of the anchor going in. He rolled off the boat when we reached the top float and I passed him his extremely heavy camera. The water, which looks flat black from aboard ship, was gin clear—alluring, if only it weren't a tick or two above freezing. He had to swim hard to keep up with the float. It looked like work. We all respect the guy for his intrepid work and the Skip for his willingness to help him do it.

This cruise has attracted unusual attention from outside the professional community. There have been reports from Reuters and other public media in addition to *National Geographic*. This is in part because there is an element of exploration to it. Once the moorings have been laid across the Denmark Strait, we'll strike off in search of the origins of that newly discovered current, which the co-PIs believe delivers half of the water flowing into the strait. Their hypothesis is that its origin lies out to the north-northeast. But does it? And if so, how far? As I mentioned yesterday, the circulation in this region is dizzyingly complex and confusing. And as Bob reminds us, their hypothesis could be flat wrong. That's why they call it exploration.

Rob came back aboard the RHIB and we took a couple of fast spins around the ship to get some establishing shots for him and Sindre.

"Captain, when we get close to Iceland..."

"Yes?"

"Can I get up on an iceberg?" Rob asked.

"Uh, that's a no."

We laughed and did a couple of riffs on how the Skip would explain it to his employers ashore if Rob came to grief. "Now let me get this straight, Captain Sheasley. You put a visitor on an…iceberg?"

"Do you know anything about farming, Skip?"

Chief Mate Adam Seamans called him from the bridge on the VHF. "We're five minutes to station."

So we headed for the stern, lined with waiting spectators, to shoot the anchor deployment. The mooring streamed back from the ship 750 meters to the top float. Rob rolled into the water and hung onto a line from the transom as the ship dragged him along at a full knot. I passed the big camera down to him. It must have felt like an anvil in the rush of water. We followed in the boat.

"One hundred meters from station," Kyle the bosun called down to us.

They hoisted the 2,200-pound anchor off the deck and ran it out over the water. Rob was getting tired hanging on. Who wouldn't?

"Fifty meters," Kyle called.

"Must be the longest hundred meters of his life," the Skip observed.

Cameras went up before every face on the transom.

Additional flotation to keep the wire vertical was rigged just above the acoustic release in the form of basketball-sized glass spheres covered with stout yellow plastic jackets. Called "hardhats," there were twelve of them bolted to the wire.

It seemed to take forever before… "We're here."

John yanked the trip line and the anchor vanished. The hardhats plunged down after it in a burst of spray. In the distance, we could see the big top float waterskiing on the surface back toward the ship before it was dragged under.

Rob was half-frozen. He said he was feeling chilblains after he climbed aboard and we headed back to the starboard side for pickup.

That evening about eight of us stood at the portside rail like tourists to revel in a fine sunset shining on low, light-gray clouds above a white bank of fog. There's no place like the high northern latitudes and at that moment no place we'd rather have been. We said so more than once before we got too cold to remain topside.

Top: National Geographic videographer Rob McGregor in position for the anchor drop.
Bottom: Coming back aboard—stiffly. Water temp: 35° F.

TEAMWORK

Watching the mooring operations, I was thinking about the unique combination of heavy industry, advanced science, and demanding seamanship that characterizes at-sea oceanography.

Opposite: A pale sun trying to penetrate arctic fog.

Conditions today are decidedly different from yesterday, alas. Visibility has shrunk to a ship's length in leaden fog with only a bone-white glow where the sun is supposed to be. Again, the shipboard mood matches the weather, only in the opposite way from yesterday. People are going about their work as usual but with a quiet and subdued aspect, speaking softly about the business at hand. Moderate wind from the southwest sets *Knorr* rolling languidly, a motion that seems also to match today's mood.

Watching the mooring operations, I was thinking about the unique combination of heavy industry, advanced science, and demanding seamanship that characterizes at-sea oceanography. If you venture aft onto the transom without a hard hat and life vest while they're laying moorings, Kyle the bosun will peer at you reproachfully over his sunglasses, tap his head, and point forward. This regulation is inviolate for good reason. Cranes are swinging very heavy objects overhead, lines and wires are stretched under load, heavy-duty winches are turning, and there's always something behind you to fall over. And that's in fine weather. Add a pitching deck when big seas are running under her stern and the jeopardy level, not to mention the requisite skill level, soars.

Then you come off the deck into the main lab and there's Bob (U.S.), Kjetil (Norway), Laura (Netherlands), and Hédinn (Iceland), all four of them world-class scientists, analyzing fresh data and composing cross-section profiles of the local ocean using special-purpose computer programs while surrounded by esoteric, fine-tolerance measuring devices lashed to the tables and floor waiting to go overboard. All four scientists have contributed moorings (and brain power) to the project, but Laura and Hédinn will leave the ship next week when mooring ops are complete. Then Bob and Kjetil will lead the exploring expedition in search of the headwaters of the North Icelandic Jet, requiring the

Chief Scientist Bob Pickart.

Top: Launching the CTD, the primary tool for determining essential physical properties of seawater.

Right and bottom: Meanwhile, mooring work proceeds.

Retrieving the CTD package.

quick and nimble analysis of real-time data to decide where to go next, an atypical sort of oceanography. They're all working now at their computers in what we call a lab, but this is not the sort of white-coated, controlled-conditions work evoked by the word *laboratory*. This one moves constantly. (All our computers are lashed to the desks; sometimes you need to hold on with one hand and type with the other.)

You can always tell which scientist's mooring is going in the water and when. They are all normally open, funny, and relaxed, but not when their mooring is being assembled and deployed. Then, clipboard in hand, they stalk from the lab to the deck and back with head-down determination, tense in abstracted concentration. Though they know consciously and appreciate that they're in the best of hands with WHOI mooring techs John Kemp and Jim Ryder handling their equipment, it doesn't matter. This is a big thing for them. They've purchased the mooring components with grant money from the National Science Foundation or its European equivalent and shipped the stuff to Iceland from their respective institutions to be loaded aboard *Knorr*. Have they forgotten something? Have things been lost in transit? Is everything working? Will the acoustic releases actually release? But there's more to it than just the equipment.

On the transom, ready to launch a mooring.

Lena Schulze easing away on the "air tugger."

There's the data. They've planned at least part of their immediate futures around its retrieval and analysis. The data is often the linchpin of a long, ongoing project. They intend to write and publish papers that depend on the ocean data the moorings and the other devices are meant to collect. The data is more valuable than the instruments.

But there wouldn't be any data, there wouldn't be any physical oceanography without this ship. We'll talk as the days go by about *Knorr* and the various shipboard departments (engineering, galley, deck, and bridge). But for now, we need to credit the unity formed by the technicians, the scientists, and the seamen aboard *Knorr*. Absent one, the others couldn't succeed. However, if visitors from off the street came aboard for a week and observed the entire operation, the unity, they could be excused for thinking, Well, yeah, there's a lot of stuff going on, but all in all it's pretty easy. That would be because all three components of the unity make the complex, difficult, physically and intellectually demanding work *look* easy.

Tomorrow, word is, we'll come close aboard the east coast of Greenland at about 68° North. It is the most spectacular coastline this side of Antarctica, as the photographs will illustrate—if the fog lifts. If it doesn't, you'll have to take my word for it. Or not.

GREENLAND

While I hate to admit it, some landscapes are so spectacular they're revealed more evocatively by photography than by language. The east coast of Greenland is one. However, yesterday morning it appeared that neither medium would be efficacious. We steamed toward the coast in "thick o' fog." You could see the bow from the bridge, but not clearly. A big iceberg appeared on radar barely a mile off the starboard bow long before its ghostly presence hove into sight. Another off the port beam remained invisible except by radar. The fog thickened.

"Well," someone said with a shrug, about the third to do so, "we'll have two more chances later in the trip." There was scientific purpose to our close approach to Greenland, but another objective was to acquire footage for the *National Geographic* TV special. So the deck crew craned the RHIB (a small motorboat) over the side. Rob, the *Nat Geo* videographer, Rachel, our still photographer, and Chief Mate Adam clambered aboard and sped off after the nearer iceberg to get whatever images the murk allowed. Hopeful members of the science staff and off-duty crew stood around the main deck, cameras in hand, but disappointment was beginning to sink in.

Captain Sheasley brought *Knorr* to a stop, waiting. A glint of sun revealed a patch of blue overhead.

"It's clearing!"

No, ten minutes later the woolly fog closed the hole.

"*Knorr.* Avon," Adam called on the radio. (Avon is the manufacturer of the RHIB, and so it became the call sign.)

"Go ahead, Adam," said the Skip.

"Rob wants you to bring the ship over so he can shoot her against the berg."

"Okay, Adam." He turned to me, hanging around on the bridge. "You want to drive?"

They can drift around for decades, depending on the currents, but most East Greenland icebergs are sent south into the shipping lanes by the East Greenland Current.

Opposite: A grounded berg off the east coast of Greenland with the inevitable kittiwakes riding the crest.

"Sure," I said.

"Oh, God," said my "friend," Pete the chief engineer. (Pete's Polish; we call him "Chiefski." Among other things.) "Now, you see that berg, right?"

"The big white thing?"

"Right. Don't hit it."

"Steer one-five-zero," said the Skip.

"One-five-zero." There were other "witty" bits of irony. "Close all watertight doors," somebody said, but I ignored the rabble. Peasants.

The berg, about the size of a toppled midtown-Manhattan office tower, took form as we drew nearer, hundreds of kittiwakes perching on its crest.

"Avon. *Knorr.* What's Rob want us to do?"

"Circle the berg," Adam replied.

This was an old berg—they can drift around for decades, depending on the currents, but most East Greenland icebergs are sent south into the shipping lanes by the East Greenland Current—veined with cross-hatched strips of blue ice and evidence of previous waterlines formed when its center of buoyancy changed as it melted. (See photos, better than words.) You don't want to approach too close to these things lest you get swamped when they calve or topple.

While we were waiting, the CTD crew did another cast. I've been waiting to discuss the CTD, its purpose and mode of deployment until after the mooring ops are complete. Suffice it to say for now that this elaborate device, fundamental to physical oceanography, measures temperature, salinity, and current velocity. That's essentially the same kind of data collected by moorings, but with the major difference that the CTD is portable. Once the mooring operations are complete, Bob and Kjetil—using the CTD as their main tool—will

Horizontal wisps of fog, as if called up for the sake of drama, hung in the air halfway up the mountains while waves broke around their base.

Cape Tupinier, 68°40′N by 26°20W.

Knorr waiting for the photographers' return.

lead us in that search for the headwaters of the North Icelandic Jet. We'll talk CTDs at that point.

"Look! It's clearing."

Come on, we'd heard that before. No, this time it really *was* clearing. We could see that magical coastline off the port side. Enough watchers and photographers scurried over to induce a port list. Thousand-meter saw-toothed, naked-rock cliffs skirted by scree fields and talus slopes soared vertically from sea level. Horizontal wisps of fog, as if called up for the sake of drama, hung in the air halfway up the mountains while waves broke around their base. Layers of pointed peaks, one higher than the next, stepped inland toward the great Greenland Ice Sheet. New sun glinted theatrically on the icebergs. You have to remind yourself that the East Greenland coast is real, not the product of some scene painter's over-caffeinated imagination. If I turn away, will it still be there when I look back?

This was Cape Tupinier (68°40' North by 26°20' West) on the Blosseville Coast, a couple of hundred miles south of Scoresby Sound (Kangertittuaq, in Greenlandic), the largest fjord in the world. There remain few unexplored places on Earth, but the Blosseville coast may well be one. Here there is almost no flat land if you don't count glaciers coursing down to the sea from the Greenland Ice Sheet, nothing to sustain human travel—or life. Not even the Inuit come here.

The Skip called the Avon back to the ship to change crews. Laura wanted to grab a sample of inshore water. She's studying the East Greenland Current (EGC) flowing southward out of the High Arctic along the shelf into the Denmark Strait. The EGC contains a relatively high quantity of freshwater from local melting, Siberian river runoff, and inflow from the Pacific Ocean through the Bering Strait. By repeatedly measuring its oxygen isotopes, she can begin to determine whether the freshwater content is increasing. Too much freshwater on the surface will inhibit the vital convection (sinking) that occurs north of the

Kittiwakes on the grounded berg.

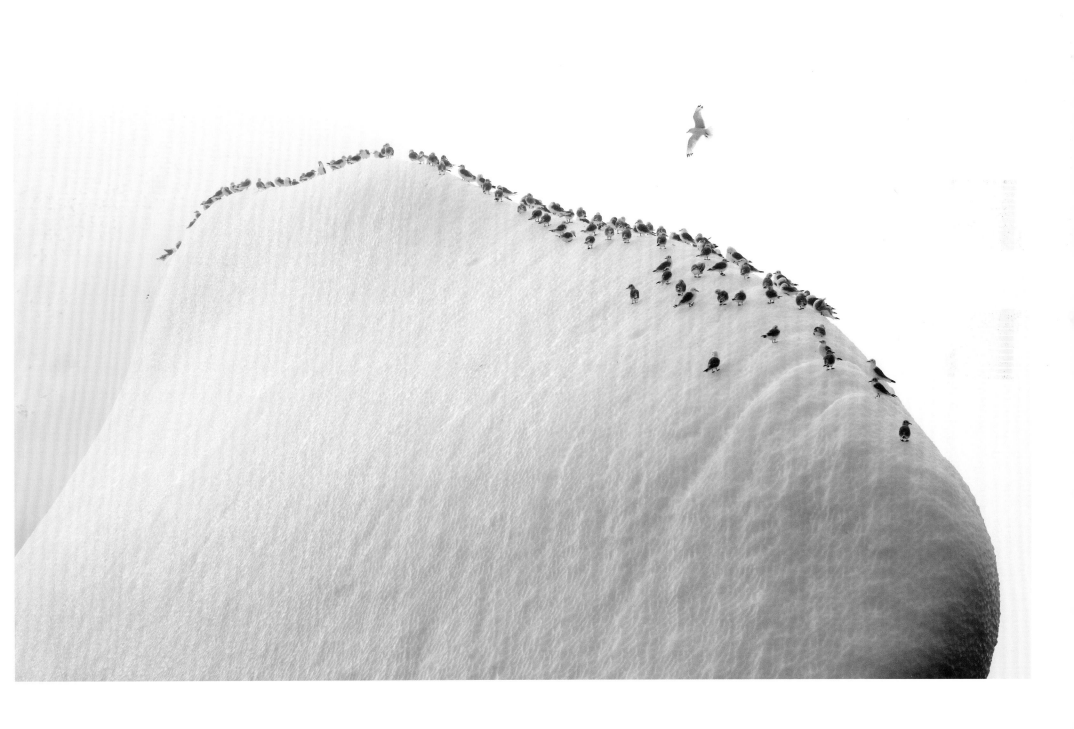

One of the many glaciers on the Blosseville Coast snaking down from the Greenland Ice Sheet.

Denmark Strait. However, no one is yet certain how much freshwater is "too much."

The Avon headed back toward the coast and soon disappeared behind a line of bergs close enough to the shore to be aground. The plan, for photography's sake, was for us to bring *Knorr* as close inshore as the Skip dared. He took back the helm—"I might have to take evasive maneuvers." We proceeded toward the cliffs at two knots, eyes on the fathometer—it never climbed above 65 meters—to within a couple of miles of the "beach."

We searched for the boat with binoculars. The radar wasn't of much use with so many targets in the water.

"Avon. *Knorr*. Where are you, Adam?"

"Between the set of bergs on your right."

"Okay, we have you." The Skip parked his ship in a photogenic position.

And from there we ogled the magnificence for the next two hours until, inevitably, we had to get underway for the final three mooring sites. Too soon Greenland vanished. Maybe it was never there.

There remain few unexplored places on Earth, but the Blosseville coast may well be one. Here there is almost no flat land if you don't count glaciers coursing down to the sea from the Greenland Ice Sheet, nothing to sustain human travel—or life. Not even the Inuit come here.

ICELAND, IF BRIEFLY

The high, serrated north coast of Iceland hove into sight before breakfast. The wind was slack, seas glassy, about eight shades of gray in the clouds. As we approached within several miles, the mountains resolved themselves into individual peaks and cliffs patched with greenery and sliced by thin, silvery waterfalls. The moorings, the focus of the cruise so far, have all been deployed across the Denmark Strait ahead of schedule, thanks to the fortunate spate of fair weather. The aft deck, so recently covered by mooring balls, spools of oceanographic wire, stacked anchors, and crates of miscellaneous hardware, looks now like a dance floor. It's kind of sad, however. It means that seven of our shipmates, their work complete, are leaving us in a few hours.

Friendships form quickly on these cruises. Sometimes we keep in touch or meet again ashore, but the context—the close quarters and shared objectives of shipboard life—that accelerated friendship are absent. However, the oceanographic community is small; we'll be to sea again with some of these friends. The Icelandic oceanographers will be leaving, so too the Dutch; Jim and John, the mooring specialists, and Rob, the *National Geographic* cameraman, who has become part of the team, will be going ashore. The starboard-side deck is now lined with their luggage and equipment.

There isn't much flat land in northern Iceland to accommodate human habitation or mobility. Away off the starboard bow, we see a tiny, utterly isolated farmstead barnacled to a narrow footing at the base of the cliffs. Standing at the rail watching his country, Hédinn nodded toward the farm and said, "A hundred and fifty years ago my forebears lived there." Reading *The Fish Can Sing* by the Nobel laureate Halldór Laxness on this trip, I've grown interested in the rural life in Iceland and the dryly hilarious Icelandic sense of humor, evinced by Laxness in his book and Hédinn in person over the last ten days. A road ran between the little farm and the shore, then disappeared into a tunnel through a

A road ran between the little farm and the shore, then disappeared into a tunnel through a buttress in the cliff side.

Harbor-entrance light, Siglufjördur, Iceland.

The town of Siglufjördur, population 1,206 in 2011.

buttress in the cliff side. These tunnels, built in the 1960s, made the shore road possible, he said, and thus relaxed the isolation of life in the outlying villages. I'll miss his wit and his sense of irony.

The plan was to drop our departing shipmates in the little town of Siglufjördur at the apex of a glacier-carved notch in the coast; from there they'd travel by car to Reykjavik and then go their separate ways. Dead slow, *Knorr* nosed into the short fjord between the jaws of two near-vertical cliffs, the scent of fish in the air. Now, in late August, there were only scant remnants of snow in depressions at the higher altitudes, no ice on the cliffs or in the sea (September is the ice minimum in the Arctic). But the evidence of ancient ice is everywhere imprinted on the landscape, in cirques and arêtes and glacier-carved passes through the pointed mountains. You can almost see the Pleistocene ice carving up the landscape and plowing great portions of it into the sea.

"Kick it up 150 RPMs," said the Skip to Jose at the helm.

"One fifty," he replied.

"The jetty's in sight," said Second Mate Jen, looking through the binocs. "And there's an object in the water. It looks like a boat mooring."

Evidence of ancient ice is everywhere imprinted on the landscape, in cirques and arêtes and glacier-carved passes through the pointed mountains.

Approaching Siglufjördur.

"Let's bring down the bow thruster."

Jen phoned the engine room to request the bow thruster. *Knorr* isn't powered like a typical ship with two or more fixed propellers on the stern. Her job requires her to stop in the open ocean while scientists measure it by various means. Therefore, she's powered by twin stern-mounted "thrusters," propellers that can pivot 360 degrees to deliver thrust in any direction and any combination. Plus, she has that thruster in the bow, which when not in use is retracted into the hull. When lowered, it too can pivot in a complete circle. So if all three thrusters are aligned ao as to direct thrust toward the center of the hull, she'll remain stationary. It's not quite that simple when ocean forces, wind and current, are exerting themselves on the ship, but today in flat calm, it was easily that simple. And that was the plan for today, to use the three thrusters in lieu of an anchor. There was an ample wharf in town, but not enough water to accommodate *Knorr's* 16.5-foot draft (23 feet, with the bow thruster down). So our people were to be shuttled ashore in the RHIB.

Those of us staying aboard helped load their gear; then we hugged them good-bye. Three ferrying trips later, they were away, and the ship felt empty.

A curious harbor seal.

Below: A large jellyfish in the plankton-rich waters.
Bottom: Typical north-coast geography.

Those of us staying aboard helped our departing shipmates load their gear, then we hugged them good-bye. Three ferrying trips later, they were away, and the ship felt empty. Then the Skip offered to take whoever wished to go for boat rides around the fjord and the waterfront town—just for fun. We all appreciated that and respected him for the offer and Bob for the time. But as Catie, one of the two-member Shipboard Science Support Group (SSSG), said, "That's one reason why people go to sea in research ships."

Paul, one of the ABs, was driving the RHIB when I went for a ride, and he had a bit of nautical business to conduct: The Skip hates it if, when entering port, his ship carries the slightest list, or side-to-side imbalance. You measure list using the "plimsoll marks," white numbers painted near the bow indicating the depth between the waterline and the bottom of the hull. Paul stopped on both sides of the bow and by radio read the numbers to Kyle the bosun. The numbers are precisely six inches high, so it's possible to read the marks with a degree of accuracy. If the numbers on both sides of the ship match, then she's riding "on an even keel," just the way the Skip likes it. She was. So off we went for a delightful boat ride. That's why people go to sea on research vessels, as the lady said, and why I feel this real emotional attachment to *Knorr* and her people.

· · · · · · · · · · · · · · · · ·

We were underway by mid-afternoon bound west back toward the Greenland coast and our second chance to see the big island while conducting serious ocean science. There's a potential problem looming, however. Remnants of Hurricane Irene have formed into a black depression right on our intended track. The Danish Meteorological Institute is calling for gale-force winds, maximum 40 knots, with seas around 15 feet for Friday and Saturday. That's not enough to scare this ship, but it's easily enough to bring sea ice down from northeast Greenland, blocking both the science and the sightseeing.

By dinnertime, Iceland had vanished astern as we steamed for Greenland—and the new weather.

Back at sea, heading west.

HOW TO MEASURE AN OCEAN, PART TWO

The ship felt empty this morning. There are some advantages to a small complement, such as single rooms and streamlined scientific procedures, but it felt lonely, bereft at breakfast, more like a transit than a full-on oceanographic expedition. The fog is as thick as I've seen it all cruise, though Second Mate Jen says she's seen thicker, when the surface was invisible from the bridge. The Skip did a routine turn around the deck after breakfast to see that his ship was battened down and ready for heavy weather. The forecast calls for something around 40 knots after midnight Friday morning continuing through Saturday. When sailors say *weather*, they mean *wind*. Wind, because it breeds waves, determines the quality of shipboard life. We'll certainly notice 40 knots, but it won't seriously erode our quality of life. However, since we'll be in the thick of it when we reach East Greenland, a local atmospheric phenomenon—the barrier winds—may swing into action.

Picture a counterclockwise-circulating low-pressure system moving from the west (often out of continental Canada or up the U.S. coast) into the Iceland Sea north of the Denmark Strait. When the northwest quadrant of the system encounters the high, steep East Greenland coast (the barrier), the wind, which cannot climb the mountains, is thus channeled southward and at significantly increased velocity. It can be blowing hell by the time the winds reach the Denmark Strait. That, however, is unlikely in 40 knots. We're more concerned that the north wind will usher the ice southward into our study area, blocking access to the East Greenland Current. And so we wait for an ice report from the Danish Meteorological Institute. This, then, is an opportune time to introduce that other most fundamental oceanographic tool, the CTD.

It stands for conductivity, temperature, and depth. By measuring conductivity and applying an algorithm or two, you can determine salinity. Temperature and depth are pretty self-explanatory.

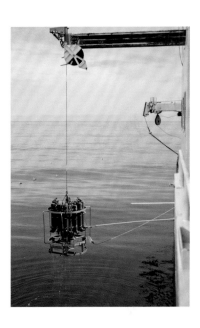

The CTD package is launched from the main deck amidships on the starboard side by a winch on the deck above. The sensors automatically log the T/S data as the CTD is lowered to within five meters of the bottom.

Opposite: Another CTD retrieval.

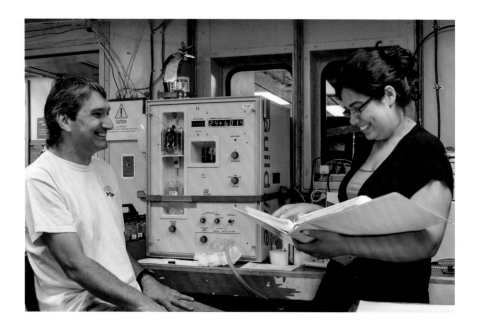

Above: WHOI technicians Dave Wellwood and Carolina Nobre enjoying work.

Above, right: Carolina, CTD-watch supervisor, with operator Mirjam Glessmer.

Right: CTD watch as seen from the 02 deck.

It's a tough schedule, interfering with normal mealtimes and sleep; everyone respects the CTD watch standers. But it looks like they're having fun, at least sometimes.

By knowing all three, you can identify the water's fingerprint and thereby follow the current in question. The device itself, weighing in at about one ton, sometimes called the "package" or, as I've heard on other ships, the "rosette," consists of a set of about twenty-four water sampling bottles like thin scuba tanks mounted on a circular metal frame. The bottles, which must be open on the descent lest they be crushed flat by the water pressure, can be triggered to close at predetermined depths from a computer in the main lab. Sensors on the CTD transmit the data through the specialized oceanographic wire to that same computer. Also, the CTD is a handy, readily used platform on which to mount other instruments such as an ADCP or biological sensors.

On this trip, we're collecting water samples in order to double-check the accuracy of the salinity sensors on the CTD by means of a device called a salinometer, operated by my lab mate, Dave Wellwood. For other applications, chemical and biological oceanographers often collect gobs of water samples to measure nutrients, oxygen, nitrogen, and other things depending on the cruise objective. On this cruise, the objective is to follow the North Icelandic Jet wherever it leads, and for that Kjetil and Bob need to know only temperature and salinity, or the T/S signature, as it's often called, and velocity as measured by the ADCP.

So that's the machine itself, but its deployment is a big production, partly but not exclusively because it's so heavy. The CTD package is launched from the main deck amidships on the starboard side by a winch on the deck above. The sensors automatically log the T/S data as the CTD is lowered to within five meters of the bottom. The CTD operators keep an eye on the computer-reported depths to be sure that the package doesn't hit the bottom. Then on the up-cast the CTD watch standers, communicating by radio with the winch operator, call for the predetermined stops to "fire" the water-sample bottles, which, with the press of a computer key, snap closed. At each stop on the up-cast, the watch manually records temperature, salinity, and depth as a backup to the automatically recorded data. Then comes the recovery process.

The winch operator hoists the package out of the water and raises it about three meters above the surface. Then, from the deck, the CTD watch standers hook lines onto the steel frame and help guide it down inboard onto a movable pallet. The cast process is not complete until the watch extracts the water samples into small, numbered bottles coinciding with the depth at which they were taken, washes the package with freshwater, and rearms the spring-loaded bottle-cap mechanism. The pallet is set on rails so that the CTD can be pulled by a winch into the adjoining hangar for its protection and that of the operators.

CTD operations are often conducted by oceanography students, sometimes their first experience of seagoing science. Work goes on around the clock. We have two three-person watches, all women, several with previous experience. They work twelve-hour watches, 1500–0300 and 0300–1500. It's a tough schedule, interfering with normal mealtimes and sleep; everyone respects them for doing it. But it looks like they're having fun, at least sometimes, and they've found a fine spirit of comradship.

Bob and Kjetil have chosen various lines across the Denmark Strait and others to the north along which they've planned numerous stops for CTD casts. We're now north of the strait (68°29' North by 022°19' West) heading northwest 333° toward Greenland. Bob is well-known—his operators might say notorious—for his closely packed, picket-fence CTD stops. But as he points out, this is the best, the only way to get high-resolution profiles of the current.

I was just out chatting with the second watch, getting in their way as they prep the package for the next cast. It's 1930 local time (that's 7:30 PM to those ashore). The wind is up around 20 knots (scientists designate wind speed in meters per second), and it's distinctly colder than it was two hours ago. I've seen CTD operations aboard *Knorr* continue until the wind reaches into the mid-thirties or when blue water is coming over the bulwarks, jeopardizing the equipment and the people. Maybe it won't come to that. In any case, we'll be right here to find out.

"Arming" the CTD package.

DOGS BEFORE THE MASTER

We began to feel the swells about midnight. Swells, technically speaking, are different from waves, though they're often mistakenly used interchangeably. Waves are directly wind-driven, whereas swells are either leftover surface disturbances after heavy weather has passed or disturbances pushed ahead of an approaching storm still some distance away. The latter was probably the case last night. The old salts in Cape Horn sailing ships used to call swells "dogs before the master." Now we're seeing genuine waves, in a tick over 30 knots, with those distinctive snakes of foam running down their faces. Three icebergs lie out on the horizon, and under the glowering gray skies, the waves breaking against their waterlines shine as if with internal light. It's really quite beautiful, though there are those who'll think me an idiot for putting it that way. But it's easy for me to say, since, lacking seamanly responsibilities, I have the leisure to observe nature.

The Danish Meteorological Institute (DMI) predicts the gale to last through today and Saturday, so the seas will likely build, since duration is one of the factors responsible for wave formation. Wind velocity is one of the others, obviously. Then there's fetch, the distance of open water over which the wind blows. As I mentioned in passing the other day, the longer the fetch, the larger the seas. Our position as of 1100 Friday (69° North by 023° West), 120 miles north of Iceland, 90 miles east of Greenland, puts us in open ocean with no land to windward. The fetch is virtually unlimited.

However, this doesn't imply, given the present wind velocity and fetch, that the seas will just build and build over the next day and a half, because another fundamental principle comes into play—the fully developed sea state. For a given velocity, fetch, and duration, waves will reach only a certain height and grow no higher. A way to say it is that the sea is filled up with energy under present

Anomalously higher waves occur because all waves don't travel at the same speed. Sometimes the "master" may overtake the "dogs" and thereby increase wave height.

Opposite: The weather changes quickly in these Arctic latitudes.

Taking water over the main deck.

conditions and can't take any more. This would change if more energy were introduced in the form of increased wind velocity. But in this, there is another *however*.

When meteorologists report wave height of, say, 10 meters, they are not talking about every wave. They're talking about the highest one-third of the waves, or in the parlance, the significant wave height. But a percentage of the waves in that one-third will be higher, sometimes much higher, than the average. The old Cape Horners used to call them "niners," observing that every ninth wave in the wave train was considerably higher than its colleagues. This may not have been very scientific, but it illustrates the principle. Anomalously higher waves occur because all waves don't travel at the same speed. Sometimes the "master" may overtake the "dogs" thus increasing wave height. And things get really miserable when a storm passes, leaving behind big

swells, and then another storm blows in from a different quadrant, setting up what's descriptively named a "confused sea." Then the poor vessel gets battered from two directions at once.

The above description applies more directly to mariners than to seagoing oceanographers, but not by much. Bob just stopped by to show me the latest ice report from the DMI. (None of us will forget their advisory from the 2008 cruise: "URGENT: Any vessels between 18 and 20 West should seek shelter IMMEDIATELY. Hurricane-force winds…") These people have been of invaluable service to the expedition, and Bob, now on a first-name basis with the forecasters, has been effusively complimentary. Their ice report is highly customized in its scale and detail, a flag marking *Knorr's* position as of 0600 UTC, the new name for Greenwich Mean Time. The report shows a 100-mile-long banner of sea ice streaming southeast from Greenland at about 75° N out into the Iceland Sea as a result of our gale-force northerly. This forestalls, for now, Bob and Kjetil's plans to complete the west end of the present CTD line, then steam north along the Greenland coast to begin a new line along 70° N. To reach that line would have required us to steam right through the swath of ice. We can't do that. *Knorr's* bows are reinforced against ice, but she is by no means an icebreaker.

Sea ice, because of its inherent threat to mariners, has earned colorful names depending on its size and concentration—grease ice, brash and frazil ice, pancake ice, and nilas, for instance. But the most dangerous is an often-solitary chunk of glacial ice less than one meter high and less than five in length called a growler. Growlers, unlike bergs, are too small to show up on radar. The point for us is that sometimes the concerns of mariners and those of oceanographers coincide on the surface of the sea. Today is one of those occasions.

Deploying the CTD in 35 knots of wind.

There's another way mariners and oceanographers share similar concerns and problems—mechanical and/or electrical breakdowns. The all-important sensors in the CTD have died due to flooding. The techs were poised to swap out the entire sensor package for a spare only to find that one of the new sensors didn't work; they may or may not be able to fix it. But even if they manage to get it functioning, we'll likely divert to Iceland for a replacement package. I doubt Bob would dare continue without a spare, given that the rest of the cruise depends exclusively on CTD data. All this high-tech electronic gear that makes modern physical oceanography possible is wonderful…when it works.

It's rougher out here than it looks.

SAVED BY ANTON

Anton Zafereo, one of the two Shipboard Science Support Group (SSSG), worked through the night (with Catie Garver, the other SSSG, right beside him) to repair the flooded pressure sensor that records the depth at which the CTD is logging temperature and salinity data. That data is useless if there's no accurate depth to accompany it. None but the most senior shoreside technicians at Sea-Bird, the manufacturer, ever disassemble the pressure sensor for service. Anton talked with Sea-Bird experts by sat phone; they warned him that if he disassembled the unit, he might never get it back together in working condition. But there was no choice and nothing to lose, so Anton went at it.

A spare sensor package is being flown to Akureyri, Iceland's second largest town, on the north coast, but it won't arrive until Tuesday. We're going to go get it in any case—it's too risky to be without a spare for obvious reasons—but had Anton been unable to fix this one, science operations would have been crippled, three precious days lost, gone. A knot of people stood tensely around the CTD computer console this morning watching the progress of the first post-repair cast. The package passed 900 meters. The pressure sensor was working flawlessly. Now instead of losing three days, Bob and Kjetil can complete this line, which ends near Greenland, then pick up another line north of Iceland on the way to Akureyri.

We've seen Bob remain cool in the face of oceanographic fire before, when heavy weather blew out an entire week, when gear failed or was stolen by the subject of his study. It happens to everyone who studies the sea at sea, but perhaps not everyone takes it so evenly. And he's genuinely and vocally appreciative when rescued from loss of time and gear by crew and techs. This includes

When the northerly winds blow down the mountainous coast of East Greenland, the water tries to bend rightward, that is, inshore. But it can't, at least not for long, because the coast is in the way.

Opposite: Sunset over the Greenland Ice Sheet.

Anton working through the night while *Knorr* continues to roll.

the CTD watches who've been deploying and recovering the package while boarding seas slosh around their legs. He's the paradigmatic seagoing oceanographer, in my opinion, and I'm not alone in that view aboard this boat.

. . . .

This blow, remnants of Hurricane Irene, is doing just as forecast—hanging around all day yesterday and today, climbing periodically into the high forties from the mean wind in the mid-thirties, and it's making for a really nasty ride. When we steam due west between CTD stops, *Knorr* takes on a corkscrew motion as the wind honks out of the northeast, pressing waves under the starboard side of her transom. They call it the "quarter-sea boogie." (I'm holding on with one hand and typing with the other; I'm not alone in that, either.) Perhaps you remember my mentioning those barrier winds created when a cyclonic low encounters the high coast of Greenland and the mountains act to accelerate the wind velocity. Bob just told me we're experiencing that very phenomenon right now. How he knows that reveals another fascinating physical act of nature.

"I'm recording downwelling," he said.

When wind blows over the surface of the sea, hauling water along with it, the force of Earth's rotation bends the flow to the right in the Northern Hemisphere (to the left in the other one). So when the northerly winds blow down the mountainous coast of East Greenland, the water tries to bend rightward, that is, inshore. But it can't, because the coast is in the way. However, the water has to go somewhere. So it flows down and then offshore along the bottom. Bob has seen the fingerprint in the CTD data of cold, fresh water from the East Greenland Current, which normally remains on the Greenland shelf, out in deeper water well away from the shelf. This is downwelling. Without the barrier winds, the downwelling would not happen.

Okay, I've had it for today. This motion is getting to me. Things are falling and bouncing around the boat as if with malicious intent. My eyes are spinning, along with my stomach. Good night.

Knorr approaching the great barrier where the wind is accelerated twofold.

EXPLORING OCEANS

When last we talked about physical oceanography, specifically in "Imagining *Oceans*," we introduced the Meridional Overturning Circulation. It sounds pretty daunting in that language. "Meridional" simply means along meridians, or lines of longitude, which run north-south, as opposed to lines of latitude, which run east-west. (The notion of gridding the earth in this way in order to determine geographic position was first conceived by Ptolemy of Alexandria in the second century.) In the North Atlantic, generally speaking, warm, salty water from the Gulf Stream System flows north into the Arctic and cold water returns south back toward the tropics—along meridians. In order to sustain the circulation, remember, the tropical-origin water must first sink in the Nordic Seas north of that submarine mountain range called the Greenland-Scotland Ridge—the overturning—and then make its way back south through gaps in the ridge. The largest of these gaps is in the Denmark Strait. That quick review brings us up to date. Now let's take a more detailed view of the circulation north of the Greenland-Scotland Ridge in order, later, to focus more finely on the North Icelandic Jet.

Opposite and above:
Off east Greenland in better weather.

The North Atlantic Current (NAC), flowing on the surface, has crossed the ridge and proceeded up the Norwegian coast, sharing its warmth with the land. Its name has changed to the Norwegian Atlantic Current, but nature doesn't care about that. The NAC is still transporting warm, salty Gulf Stream water. Some portion of that water—still identifiable by its temperature/salinity profile—nips over the top of Norway into the Barents Sea, and another arm of the Norwegian Atlantic Current continues on northward into the High Arctic beneath the pack ice. But a significant volume of similar water takes a great loop around the Nordic Seas basin—currents like to hug the edges of ocean basins—and proceeds south along the coast of Greenland aboard the East Greenland Current into the Denmark Strait. For us, it's important to remember that, while this is still identifiable as Atlantic-origin water, it has

been chilled by the cold ambient air, or in the parlance, it has become "densified." A large quantity of densified water, some 100 to 300 meters beneath the surface, travels with the East Greenland Current toward and into the Denmark Strait after taking its lap around the Nordic Seas basin.

In 1996, Cecilie Mauritzen, while still a graduate student at MIT/WHOI, published a paper in the technical journal *Deep-Sea Research* in which she proposed the above scenario. It flew in the face of the previous paradigm, which held that the Atlantic-origin water, having been densified by exposure to the cold Arctic air, sank at two or more locations in the interior of the Nordic Seas, then made its way southward to the Denmark Strait, where it spilled over the sill into the deeper North Atlantic. In this we have an interesting example of how oceanography and, by extension, science works. An idea is proposed and, if satisfactorily proven within the limits of known data, becomes over time the accepted paradigm.

Summer weather in these high latitudes tends to be gentler than at Cape Farewell at the southern tip of Greenland.

Nordic Seas Circulation

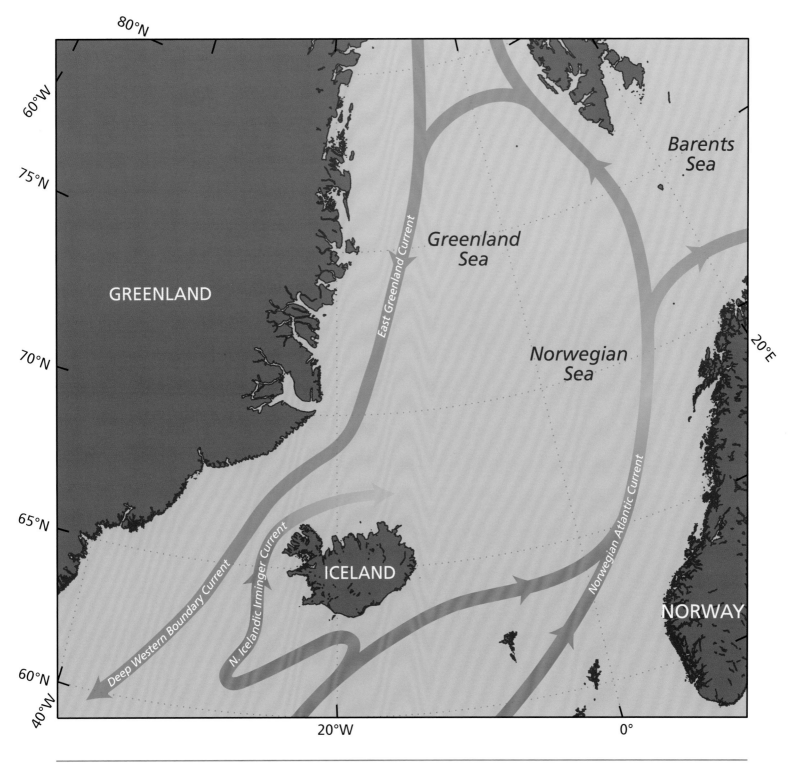

The previously accepted paradigm: Much of the warm-to-cold conversion in the MOC happens along the perimeter of the Nordic Seas.

Jack Cook (WHOI)

Then another scientist(s), perhaps armed with new instruments producing better data or a different interpretation of existing data, hypothesizes a new paradigm. (Think of Newton, Darwin, and Einstein, who revolutionized science and with it our view of the physical world.) Thus science advances, though the scientist whose paradigm is exploded may not like it.

A mere two weeks ago at this writing, Kjetil Våge, Bob Pickart, Hédinn Valdimarsson, Steingrímur Jónsson—co-principal investigators on this expedition—Michael Spall, and other authors published a paper in *Nature Geoscience* hypothesizing that fully half the water flowing into the Denmark Strait is transported there by that newly discovered current they named the North Icelandic Jet (NIJ). This does not categorically contradict Mauritzen's Denmark Strait scenario, but most definitely modifies it—that is, if the new hypothesis is accurate. The exclusive purpose of this second half of the expedition—led by Bob and Kjetil—is all about proving or disproving their own hypothesis. It isn't often that new currents are discovered. It's likewise rare for an oceanographic cruise to so closely resemble exploration in the traditional sense.

Since the days of Henry the Navigator in the early 15th century, no explorer, or at least none we've ever heard of, sailed into the unknown with no specific purpose, just to see what he happened to see. All had an a priori idea, an objective that their expedition was planned to test. One difference, however, is that the people who stayed ashore, from Queen Isabella to the Royal Geographic Society, had to wait until the expedition returned years later to learn whether the explorer found what he was

Minkes, a common whale in these waters, patrol the plankton-rich shelf waters.

looking for or something different or nothing at all. Here aboard *Knorr*, Bob and Kjetil are conducting real-time oceanographic exploration. Their tools are minimal—the CTD and ADCP. Finding North Icelandic Jet water where they expect it to be will go a long way toward proving their hypothesis. Likewise, they will need to look for NIJ water where it *might* be in order to disprove (they hope) other possible explanations for the NIJ and its origins.

In the days to come, I'll try with their help to explain the physical details of the oceanographic logic that led to their hypothesis in the first place—and the process by which they seek to verify it. The exciting thing about this cruise is that we won't need to wait the usual year or more for the data to be analyzed to learn the results of their exploration. We'll be right here when the answer comes in.

We have an interesting example of how oceanography and, by extension, science works. An idea is proposed and, if satisfactorily proven within the limits of known data, becomes over time the accepted paradigm.

THE NORTH ICELANDIC JET

In 1999, the Icelandic oceanographers Steingrímur Jónsson and Hédinn Valdimarsson, monitoring with ADCPs the slope waters close to the north coast of their home island, noticed a margin of enhanced flow in 600 meters of water. It looked very like a current. But there was no record of such a current in these waters; nobody had ever heard of one. Could this actually be a newly discovered current? Well, maybe yes, maybe no. But prudent scientists don't step ashore claiming like a conquistador to have discovered a hitherto unknown current. This may not have been a current at all, but some tidal phenomenon or a short-term variation in something or other. So they went back for another look.

And there it was, flowing from east to west toward the Denmark Strait. It wasn't enormous, transporting about one million cubic meters of cold water per second, compared, say, to the 35 million in the North Atlantic Current carrying warm Atlantic-origin water into the Nordic Seas. (By the way, ocean scientists measure water transport in units called Sverdrups, in honor of the brilliant Norwegian oceanographer Harald Sverdrup, 1888–1957. One Sverdrup, or Sv, equals a volumetric transport of one million cubic meters per second.) Still, one Sv is far from trivial. Then, in 2004, Jónsson and Valdimarsson published a paper in a technical journal read only by scientists, *Geophysical Research Letters*, proposing that their newly discovered, yet unnamed current contributes a "major" portion of the water entering the Denmark Strait.

This was very intriguing, but it was based mainly on velocity measurements with little hydrographic information, that is to say, minimal temperature and salinity data to identify the fingerprint. The Icelandic scientists of course recognized the relevance of hydrography, but they lacked the ship time to focus on the North Icelandic Jet (NIJ). And therein lies another reason why oceanography is such

Mirjam at the CTD console.

Opposite: Knorr with the ever-present fulmars. They followed us every day. When we stopped, they stopped. When weather blew in, they disappeared. When it cleared, they reappeared.

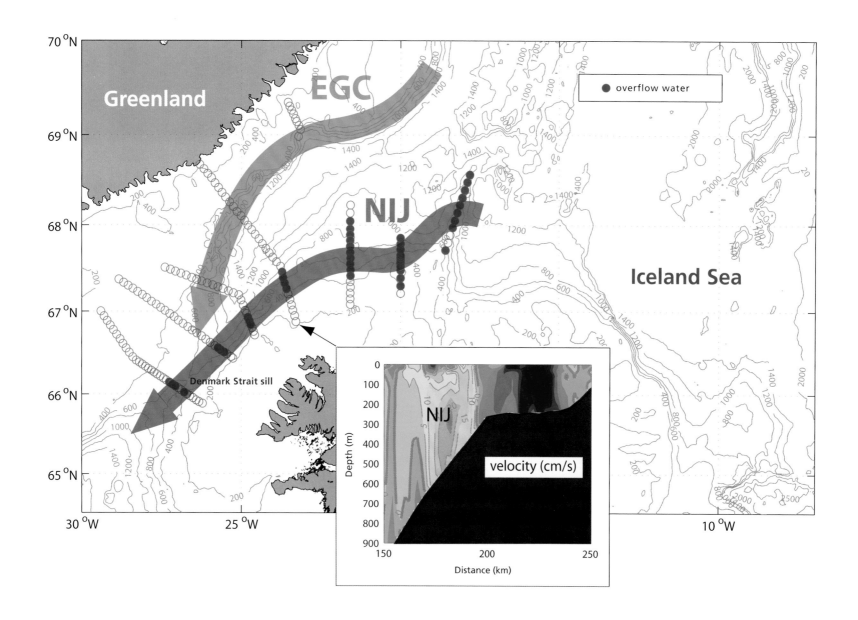

October 2008
Knorr Hydrographic Stations

Oceanographic Renderings of the North Icelandic Jet, the Objective of Our Expedition

The East Greenland Current is pictured in magenta. The hitherto uncharted North Icelandic Jet is in purple. Red circles denote CTD sites;

green circles represent water with the NIJ fingerprint. But where does the NIJ come from? Why does it exist?

a young science and why parts of the world ocean remain under-sampled. Physical oceanography is expensive, and since there's no pending profit in it, private enterprise doesn't participate. Funding must come from governmental organizations such as the National Science Foundation in the U.S. and its counterparts abroad. This ship runs about $45,000 a day just in operational costs, never mind science costs. Still, there was this sustained flow on the order of one Sverdrup. It had to come from somewhere, to go somewhere, and it certainly appeared to be heading for the Denmark Strait. If so, it had the potential to be a paradigm shifter. But at that point, this was a big *if*.

Skipping a few details, this brings us to Bob's 2008 cruise, which has, for its severe weather, gained legendary status aboard *Knorr*. The basic objective of the cruise was to pull several moorings south of the strait, replace their batteries, extract their data, and redeploy them ("turnaround" in the parlance). But between storms, Bob slipped in a few days to run CTD sections athwart the proposed current to verify its existence. Sure enough, it was right there where the Icelanders claimed, hugging the Iceland shelf and flowing from east to west. Hmm. That introduced multiple questions: Where did it come from? What caused a current in that area, and not just any old area, but one in proximity to and perhaps heading toward the critical Denmark Strait? If indeed it contributed water to the inflow, then it was likely contributing to the overflow. That alone made it important as well as merely interesting. But... a new current? More data was necessary before the old paradigm fell to the new.

It looked very like a current. But there was no record of such a current in these waters; nobody had ever heard of one. Could this actually be a new current? And there it was, flowing from east to west toward the Denmark Strait.

The Icelandic ship R/V *Bjarni Samundsson*.

Bob, Kjetil, Steingrímur, and Hédinn returned to the region aboard the R/V *Bjarni Samundsson* in 2009 to run more hydrographic and velocity lines across the current. Then everyone went home to analyze the data. Yes, the current actually did flow around the northwest corner of Iceland, no question about it. Its existence a solid fact, it now needed a name. In exploration, the privilege of naming a new feature traditionally falls to the discoverer(s). The Icelanders settled on the Northwest Icelandic Jet, later modified to the North Icelandic Jet.

And that brings us to today aboard *Knorr* on this calm, foggy morning about 100 miles off the north coast of Iceland—and to the hypothesis as to the origins of the jet Bob and Kjetil mean to test on this cruise. I have as of today only a fingertip grip on the oceanographic dynamics that "cause" the current. With help from these two scientists I'll strengthen my grip firmly enough to explain it in accessible language. (I see them now in the main lab sitting shoulder to shoulder peering and occasionally pointing at Bob's computer screen as they discuss what they're seeing, so I'll wait to corner and question them later.) But as to their on-board process, I can say this now: They contend that the jet is formed right there off the north coast of Iceland by a particular series of physical interactions between warm water and cold, light water and dense. Bob and Kjetil freely admit, however, that the current could be coming from elsewhere. Perhaps from the north, being steered southward by a submarine

The intrepid watch wrangles the CTD package overboard for still another cast.

mountain range called the Kolbeinsey Ridge. This is a reasonable possibility. Currents are often steered by underwater topography, and the ridge is well positioned to steer a south-setting current onto the Iceland shelf. It's possible also that the current flows from around the east side of Iceland. And so they will first parse, with tightly packed, high-resolution CTD casts, the slope-water region of Iceland where they know the NIJ is present—and then go search the waters north and east, where they hope it is not.

• • • • • • • • • • • • • • • • •

We're due for some more heavy weather early tomorrow and Friday, which surprises no one. In the wee hours tomorrow, we'll complete the present CTD line across the NIJ, then head to Akureyri to pick up the new CTD innards. Then it's right back out on the trail of the North Icelandic Jet.

The CTD instrument used to find the North Icelandic Jet.

ERRANDS (WET)

Opposite: Kolbeinsey Rock is the only piece of Iceland north of the Arctic Circle.

We were pounding back to Knorr aboard the little RHIB (they call it the work boat to distinguish it from the slightly smaller rescue boat) in something between 35 and 40 knots of cold, ironclad wind. The Skip was driving, Jen sitting behind him on the center-console seat; Russ the electrician ducked bullets of spray in the bow. Chief Steward Bobbie (chef, in land lingo, and an excellent one) and I in the stern clutched for handholds. We were all hunched into our foul-weather-gear jackets against the rain and flying spray. The bow surged up over each closely packed crest, then plunged repeatedly into the troughs, throwing up blasts of spray for the wind to hurl in our faces, liquid bullets. Bobbie, not frightened but not comfortable either, hugging her purse in a plastic bag to her stomach, leaned toward me and said, "I don't do this very often. Actually, never."

. . . .

This morning's coffee klatsch on the bridge was unusually well attended. The difference was *land.* We were approaching it, namely, Iceland, specifically, Akureyri, in order to pick up the two new CTDs waiting for us on the wharf.

To be geographically precise, we'd had a granular glimpse of Iceland the day before yesterday in the form of a tiny, isolated chunk of black rock half a dozen people couldn't sit on at once called Kolbeinsey. It seems, according to the Icelandic Sagas, that one Kolbein Sigmundsson from Kolbeinsdal got into some kind of 10th-century scrape and fled for Greenland. But he didn't get far, only 60 miles, before he wrecked and died on the island that bears his name. Back then, it was an island, a dry-land protrusion of some size atop the Mid-Atlantic Ridge, but passing time and ocean erosion have reduced it to a cleft rock. In 1984, it was still large enough to support a helipad, but the ocean claimed it along with most of the rock in 2006.

Kolbeinsey Rock

We watched an outbound vessel maybe 100 feet long pass us to port pitching like a bathtub toy, burying her bow in every wave, stern heaving skyward.

Bob and his post-doc student Donglai Gong, who hand-carried new CTD sensors from the U.S.

Heading back to the ship after our wet shopping trip.

As Iceland itself hove into sight, a high, hulking mass of gray slightly darker than the shroud of clouds between us and it, the bridge was crowded with gawkers, talkers, and superfluous personnel, me among them. This piece of land had special meaning to some of us. In 2008, we hunkered for four days behind a little island called Hrisey halfway down the fjord called Eyjafjordhur, while a hurricane-force beast tore up the Nordic Seas. Along the Greenland barrier it blew 100 knots; it blew 50 in our hiding place, but the seas were flat. It was October, a fresh snow had fallen; Hrisey and its sweet little town of Saltnes looked like a Christmas-card fantasy, Currier and Ives in the subarctic. So we wanted to see it again on the way to Akureyri, situated at the foot of the fjord. Meanwhile, we told each other "remember—?" stories, which those who weren't there had heard six times already.

Trouble was, we couldn't see much of the fjord for fog, rain, and low cloud. We were, however, going in the best direction for ease of motion, dead downwind in the northerly. We watched an outbound vessel maybe 100 feet long pass us to port pitching like a bathtub toy, burying her bow in every wave, stern heaving skyward.

"That'll be us in a few hours."

"Let's spend the night in the fjord," Jen suggested, though she knew we wouldn't.

The Skip rolled his eyes. Bob wasn't about to waste a night of CTDs, but we could talk. Talk is cheap.

"Chiefski, can't you break something?"

Chiefski likewise rolled his eyes, but we still did a five-minute riff on all the things that could break, nothing serious enough to really interfere with the science, you know, just something to keep us in the fjord overnight as opposed to beating our brains out pounding to windward toward the next CTD line. Both of the new sensors had already arrived in Akureyri, one hand-carried from Woods Hole by Bob's post-doc Donglai Gong, the other flown from Seattle to Reykjavik and trucked to Akureyri, so there was no hope for that sort of delay. The seas flattened a little as we cleared the jaws of the fjord. Jen's track line on the chart plotter showed that we'd pass our island to port. Gradually, it solidified as we drew nearer, then Saltnes hove into sight. It felt like we'd never left, the place imprinted on our

Chief steward Bobbie, shopping in Akureyri, Iceland.

Bottom, right: Loading fresh fruit and veggies for return to *Knorr.*

eyes. "Do these guys know something we don't about the weather, or are they just weird tourists?" A couple crossed to the other side of the street as we approached.

"Knorr" was stenciled in black block letters on the back of our life jackets, and I wondered if maybe the locals thought that might have something to do with the Department of Correction. "We're off the research vessel out in the harbor," Russ explained to the store clerks and everyone we asked for directions. Most stared at us flat-eyed and slowly nodded, ready to bolt at any sudden moves. But maybe I project.

. . . .

memories, the red-roof church, those blue buildings on the shore,…and, oh yeah, the flashing neon beer sign.

Two hours later we were parked about a quarter mile off Akureyri's town wharf.

"Want to go for a boat ride?" asked the Skip.

"Uh—" Forty knots, rain, four-foot seas. "…Sure."

"Bobbie wants to get some fresh fruit and vegetables. Other people need some things. Get your gear."

We looked like we'd crossed the Norwegian Sea in that little boat by the time we tied it abeam of the docked pilot boat and clambered ashore, where Donglai waited with the CTD sensors. Russ and I marched off in the rain toward the town center with our shopping lists and got about four blocks into town before we realized we'd forgotten to take off our life jackets. Pedestrians glanced at us out of the corners of their

The errands run, we loaded the sensors and five boxes of Bobbie's groceries aboard the RHIB, six inches of water sloshing around our feet, and bounced back to the ship, where we still had to make it up the boarding ladder without serious incident.

"I don't do this very often…."

There was no further mention of a night in the fjord, so inevitably we turned and headed north into the teeth of the wind and seas. Now we move about the ship from handhold to handhold as if walking were an unpracticed action.

THE HYPOTHESIS

Scientists seek to understand how the natural world works. Therefore, that we live in the world seems reason enough to look over their shoulders as they do so. I like that idea, though admittedly it's a bit simplistic. There are a lot of shoulders to look over. Nature's systems on land and in the air and ocean are dizzyingly complex and multiple, full of cycles, rich in oscillations, feedbacks, and variations across all timescales. And so in today's science there is no such thing as a professional generalist. Even specialists can't learn in a lifetime everything there is to know about their field. Take oceanography, itself a scientific specialization, within which there are subspecialties. Physical oceanographers, Bob and Kjetil, for instance, are concerned with the water itself, its movement. They care about the nature of the seabed only insofar as bathymetry tends to steer their currents. A geological oceanographer, however, would just as soon all that pesky water go away; it's an impediment to understanding the geology beneath it.

On this cruise, we have occasion, unique in my experience, to witness a special oceanographic process in action. It's esoteric to be sure, as it must be because of the degree of specialization required to analyze ocean dynamics, but it's not impenetrable to the rest of us. Bob just passed to me a revealing graphic depiction of the return data from the latest CTD section. Figure A (p. 76) of the three panels shows a map of the Denmark Strait, the present CTD line marked in red. Figure B (p. 77) shows the temperature/salinity profile. The third panel, figure C, showing the velocity measurements derived from ADCP data, is perhaps the clearest. See that yellow swath running down the middle of the picture stretching from the surface to the bottom at 700 meters? That's the North Icelandic Jet, exactly where it's supposed to be. This is as close to real-time oceanography as it gets. If Bob and Kjetil find the NIJ where it's supposed to be, and do not find it where it's not supposed to be, then their hypothesis

On this cruise, we have occasion, unique in my experience, to witness a special oceanographic process in action.

Opposite: CTD work goes on around the clock.

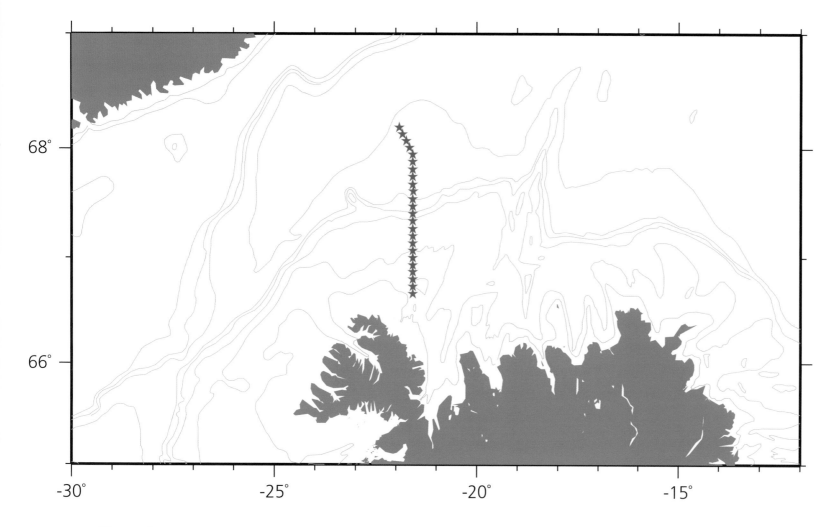

Figure A

Figures B and C show the North Icelandic Jet's "fingerprint" in the form of temperature and velocity along the transect shown in Figure A. The current is there, no question, but why is it there?

will be confirmed. That remains to be seen, but this is a good time to look at the hypothesis itself and the process by which it came to be.

First, let's state the hypothesis. The NIJ is formed in the waters on the north slope of Iceland—unlike most currents, it does *not* flow from somewhere else to arrive at the north coast of Iceland. From there and only there, the NIJ flows into the Denmark Strait, accounting for half of the water that overtops the shallow sill in the strait and spills into the much deeper North Atlantic. Or to put it differently, the NIJ is a significant part of

the headwaters of the return flow that balances the northward flow into the Nordic Seas, the Meridional Overturning Circulation. That balance also serves to moderate our climate by distributing heat.

The data from the 2008 cruise categorically verified the existence of the NIJ, and that's all. Neither Bob and Kjetil nor the Icelanders who originally discovered it knew how it came to be or why it was there. So Bob plopped the data down on Michael Spall's desk and said, essentially, "How could this be?" Spall, Bob's longtime collaborator, is an ocean modeler. Using

Figure B

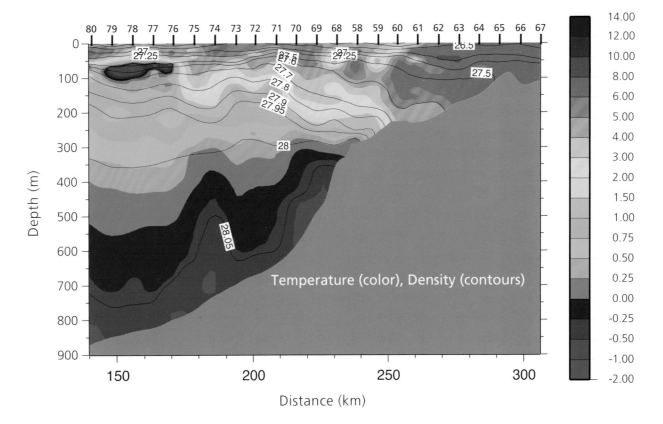

Figure C

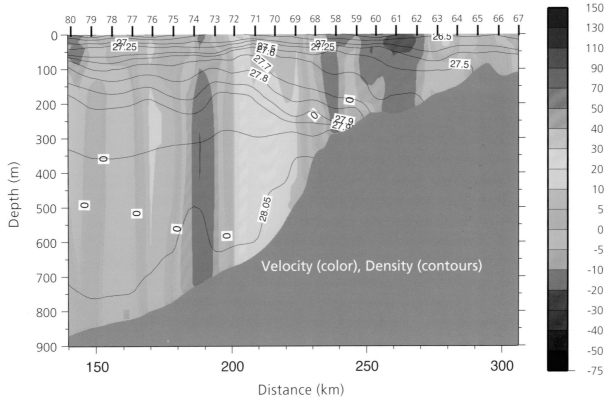

computers and higher math and a modicum of creativity, modelers create, well, a model of ocean dynamics. Modelers are a different breed of cat from observational oceanographers. (Bob tells me that when he invited Spall on this cruise, Spall replied, "What would I do on a ship?") I can't even pretend to understand how modelers do their work, except to say that they try numerically to replicate the basic laws of physics as they act on water. They can feed into the computer a simple set of parameters or a lot of detailed parameters, including but not limited to hydrography, bottom topography, and the position of dry landmasses, depending on whether they mean to produce a "simple" or a "realistic" model. They can put Iceland in the picture, for instance, watch how the water behaves, then take it out and compare the difference in behavior. Spall made a simple model, smooth topography, no realistic landmasses, because he intended to focus directly on the question at hand.

It included a current we've not yet mentioned, because no one thought it was particularly important: the North Icelandic Irminger Current. For much-needed simplicity's sake, let's call it the Irminger Current (IC). The Irminger Current is a branch of the North Atlantic Current delivering those many Sverdrups of Gulf Stream–origin water into the Nordic Seas. The IC diverges from the main flow south of Iceland and makes its way on the surface northward through the Denmark Strait hugging the Iceland side. It then flows over the north coast of Iceland at about one Sv, or one million cubic meters per second. Its existence was no mystery, but no one paid the IC much heed, because until the North Icelandic Jet came to the fore, no one

thought it mattered. But in Spall's model, considered in conjunction with the NIJ, it assumed new significance.

On the north coast, the IC plays out, devolving into eddies (swirls of water like those you see in bathtub drains, only bigger) that spin lazily out into the Iceland Sea. There they become densified as they relinquish their warmth to the cold arctic air and slip beneath the surface. Okay, hoping that's clear so far, I'm going to skirt the oceanographic technicalities and suffice it to say that what happens is a miniature version of the hemispheric process we've already discussed: If a quantity of warm water goes north, then an equal quantity of cold water must go south. Nature insists. Thus the water that becomes the North Icelandic Jet proceeds from the sinking Irminger Current water. Therefore, according to Spall's model, there could be no North Icelandic Jet without the Irminger Current. There it is, the hypothesis. It's a particular piece of scientific logic resulting from collaboration between two kinds of oceanographers, the seagoing kind, Bob, and the shoreside kind, Mike Spall.

Yeah, but what if the hypothesis is wrong? Bob is quick to say it might be. Well, then it's back to the drawing board. The existence of the NIJ is not hypothetical. It's real. If it is not formed according to the logic stated above, then it comes from somewhere else, the north, maybe, or the east. What matters is that it flows into and out of the Denmark Strait. There's no question about that. My money, however, is on these guys and their hypothesis. (Bob just warned me, however, to keep my money in my pocket.)

The hypothesis is a particular piece of scientific logic resulting from collaboration between two kinds of oceanographers, the seagoing kind, Bob, and the shoreside kind, Mike Spall.

A calm night for CTD operations.

R/V KNORR

By 1989 it became clear that, given maturing ocean science, its evolving technology, and the eclectic missions Knorr was being assigned, a bigger ship was required.

Opposite: Retrieving the boat after another photo opportunity.

I was listening to sea stories after lunch—well, literally they were land stories, about port stops—by Pete the chief engineer, Kyle the bosun, Second Mate Jen, and the Skip. Apropos of what, I don't exactly remember, but in the course of five minutes, they reminisced about shoreside, uh, activities in Namibia, Malta, and Nuuk, Greenland. They referred to the spate of nasty weather en route to Guam; or was it the Galápagos, or that time up by the Flemish Cap? *Knorr* and the nucleus of her crew have worked all the oceans of the world and most of its seas. She's been to the ends of the Earth—in the Southern Ocean near Antarctica and beyond 80° North. By 2005 she'd logged one million nautical miles of ocean science.

Owned by the U.S. Navy and known officially as an Auxiliary General-Purpose Oceanographic Research vessel, she's operated, manned, and scheduled by WHOI. Her keel was laid in 1967 in Bay City, Michigan, and, christened R/V *Knorr* after Ernest Knorr, the Navy's pioneering 19th-century cartographer, she slipped down the ways at the Defoe Shipbuilding yard a year later and proceeded to log an abiding list of impressive accomplishments.

By 1989 it became clear that, given maturing ocean science, its evolving technology, and the eclectic missions *Knorr* was being assigned, a bigger ship was required. So she was sent to the McDermott Shipyard in Amelia, Louisiana, where she was cut vertically in half and 10 meters of bottom, topside, decks, and houses were inserted. You can still see the welded seams on the main deck, starboard side, where the CTD is launched.

Most commercial ships are built for a single task—tankers, car carriers, freighters—and most steam in a straight line between Port A and Port B. However, ocean research, like the science itself, is not singular. *Knorr* needs to be uniquely versatile, to serve the geologist who wants to drill deep

Knorr in 1989 at the McDermott Shipyard in Amelia, Louisiana, where she was cut in half and lengthened to accommodate new research missions.

bottom cores, the engineer who wants to test the latest remotely operated underwater vehicle, and the physical oceanographer who wants to lay moorings across a mean body of water and do over 300 CTD casts in a month. *Knorr* routinely performs maneuvers beyond the capability of other deep-sea vessels. She stops repeatedly in the middle of the ocean, even while it's blowing 35 knots across the deck, and holds her position for as long as it takes to deploy and retrieve all manner of measuring devices. This requires a specialized propulsion system in the form of those thrusters we talked briefly about earlier. Two in the stern and one in the bow, each thruster can rotate 360 degrees (*azimuthing* is the fancy term) to direct thrust in any and all directions. She can spin in her own length, move sideways, or not move more than one meter in any direction during all but the most unreasonable weather. This of course demands equally specialized ship-handling skills and related deck work redolent of heavy industry. When I first met *Knorr* several cruises ago, I was deeply impressed not only by that skill but by the confident informality with which it was being performed. When I grew used to it, I began to assume that the high level of skill (and style) aboard *Knorr* was typical of all ocean-class research vessels. Well, it's not, I subsequently learned aboard other research vessels.

Each thruster can rotate 360 degrees (azimuthing *is the fancy term) to direct thrust in any and all directions.*

There are bergs all around, lost in the fog. But the radar sees them.

Bob seeking a moment of solitude on the stern.

Throughout history, ships and sea-boats have evoked strong, romantic emotion among their crews crossing big waters utterly on their own, their lives dependent on their shipmates' competence and their vessels' strength against the ocean's violence. It's no coincidence, given such isolation, that ships have made good literary microcosms (Coleridge, Conrad, Melville) through several centuries of Western literature. In the days of sail, even the most damp-souled literalist was stirred by images of a full-rigged ship running before a full-on gale, everything set to the royals. (By the way, *ship* had a technical meaning—a vessel of three or more masts with square sails on all masts. The word has carried over to modern, imprecise usage.) In those days, going to sea left a death-like absence at home. Maybe the sailor would return, but many did not. That isolation is less complete today, thanks to satellite communications, e-mail and Internet connections. And modern ships are not beautiful in the same obvious way as *Cutty Sark* or *Flying Cloud*. That feeling abides nonetheless. It's not hip to talk about such things, and some people might protect raw emotion with a quip or joke, but it's clear to see that many of the people aboard love this ship. I do, and I'm only a dilettante.

Utter isolation is a common condition for research vessels at the ends of the Earth.

Knorr in Rekjavik Harbor. She's been in hard waters, you can tell, by the scrapes along her waterline.

I guess I'm thinking about these abstractions because there's an element of melancholy to the story of *Knorr*. Her working life is approaching its end. Even now a new generation of research vessel is in the design stages to replace *Knorr* and her sister ship *Melville*. *Knorr* at the breaker's yard ("turned into razor blades," as they say) is too sad to contemplate—the Skip doesn't even want to talk about it—but all life, even that of ships, is ultimately replaced by the new, damnit. One hears two years, sometimes five. I hope it's the latter, but now I don't want to talk about it either. So here are some cold-eyed nautical particulars instead.

Length: 279 feet (85m)

Draft: 16.5 ft. (5m)
Bow thruster lowered: 23 ft. (7m)

Beam: 46 ft. (14m)

Gross weight: 2,518 tons

Range: 12,000 nautical miles

Speed: 11 knots cruising

Endurance: 60 days

Fuel capacity: 160,500 gallons

Propulsion: two diesel-electric stern thrusters, 1,500 HP each

Bow thruster: retractable azimuthing, 900 HP

Crew: 22

Technicians: 2

Science party: 32

LIFE'S ROUTINE

The time draws short. We'll end the trip at Isafjördur on the northeast coast of Iceland in eight days. There the science party and some of the crew will leave the ship to go their separate ways. *Knorr* will then round Cape Farewell at the southern point of Greenland and head up the west coast to Nuuk, where a new science party and their gear will board for a cruise in the Davis Strait, between Greenland and Baffin Island. And that is the nature of life aboard *Knorr*, one cruise bleeding into the next with brief port stops in between for disembarking and loading, then she's underway again.

Meanwhile, here, east of Iceland, it appears that the weather's breaking in our favor, the low finally moving off toward mainland Europe. Now, 1800 local time, the wind is still up near 40 knots, but the sun is actually shining, glinting on the whitecaps, and the gray sea has tuned teal in the low-angle light. CTD and ADCP data keep pouring in, and Bob and Kjetil continue nimbly to adjust their thinking in response. Meanwhile, shipboard routine continues with its pleasing (speaking for myself, at least), practiced regularity. This, then, might be a good time to explain the division of labor that keeps this vessel running.

This expedition is funded by a grant from the National Science Foundation. As Bob is quick to credit, multiple principal investigators (PIs) contributed equipment and brainpower during the early stages of the cruise, when moorings were the focus. Now Bob and Kjetil are directing the ship to where they need her to go in pursuit of the North Icelandic Jet. However, neither gives orders. Captain Kent Sheasley is ultimately responsible for the vessel's safety and that of all aboard. That's his first responsibility, but a close second is to facilitate science. So there is, or there should be, a close, mutually respectful working relationship among the chief scientist, the captain, and the bridge officers.

Though the chain of command is fixed, the atmosphere on Knorr *is delightfully unmilitary. No one snaps out orders. No one calls anyone sir or ma'am, at least not officially. There are no trappings of rank, nothing even close to uniforms.*

Opposite:
The wild northeast Icelandic coast.

Far left: Second Mate Jen Hickey at the helm.
Left: Co-Principal Investigator Kjetil Våge

Opposite, top: Laura de Steur, co-principal investigator.
Opposite, bottom: Adam Seamans, chief mate.

Wayne Hinkel, able-bodied seaman.

Bob is a world-class scientist, but he's also a world-class chief scientist. There's a difference. It has to do with what I think of as "scientific seamanship." He's calm in the face of adversity, he understands the way of the ship from the mariners' perspective, and he treats *Knorr*'s people with the genuine respect they deserve. (I've never sailed with a bad chief sci, but I've heard stories from the crew.) Since we're talking about the bridge, we might as well start there and see how far we get in one day.

The Skip has three mates, designated chief (Adam), second (Jen), and third (Mike). I'm not certain when the traditional roles of the mates were nominally fixed, but it reaches back well into the 18th century. Each mate stands eight-hour bridge watches in stints of six and two hours (the latter called the "dog watch" in antique usage), actually operating the ship while she's "on station" and transiting between stations or ports. That's their main job, but each has additional duties. The chief mate serves as medical officer (the Skip is also medically trained), and he's in charge of the deck, everything that goes on "outside." This includes the boats, their deployment and operation; the anchor systems; cranes and winches, and he plans the in-port loading procedure in the context of available deck space and stability. In all this deck work, the chief mate depends on the bosun, who actually gives orders to the ordinary seamen (OS) and the able-bodied seamen (AB), rather like the lieutenant-sergeant relationship.

Though the chain of command is fixed, the atmosphere on *Knorr* is delightfully unmilitary. No one snaps out orders. No one calls anyone sir or ma'am, at least not officially. There are no trappings of rank, except the size of personal cabins for officers and crew, nothing even close to uniforms. Science staff, officers, techs, crew— we all eat together in the same "mess room" and share all public spaces. Everyone knows who's in charge at the various levels and respects their capability. *Knorr's* people could make more money on commercial, unionized ships, but this egalitarian spirit is one of several reasons why they stay aboard *Knorr* trip after trip.

The second mate is, by tradition, the navigation officer. She plans and plots the routes between ports, and on this trip, she plots the CTD lines depending on where Bob and Kjetil want to go. She updates the charts per the monthly *Notice to Mariners*, and she's responsible for the bridge electronics. Bob told me a good nav and ship-handling story the other day. The latitude-longitude positions are often carried out to several decimal places. For instance, our present position is 65°12.448' North by 011°47.215' West. There are 60 nautical miles in each degree of latitude, and each minute of latitude equals one nautical mile. So we're presently 12.448 miles north of 65°

North. Early in the trip, Bob planned a mooring position, which always needs to be precise, and when we arrived on station, he went to the bridge to check where we actually were. Well, we were spot on, directly over that exact position on the bottom—out to three decimal places. Adam, who happened to be on watch, asked Bob what point on the ship he'd like to plot from, the fantail, the transom? "I don't bother going up to check positions anymore," Bob told me.

In addition to his bridge watch, Third Mate Mike is responsible for personal and ship's safety gear. We all have life jackets and survival suits ("immersion suit" is now the preferred term) in our

Off-watch time in the library.

Captain Kent Sheasley installs
the CTD waypoints list.

Third Mate Mike Chrétien.
Everything that happens aboard
has to be logged.

rooms, which Mike maintains and services along with the life rafts, fire-fighting gear, and hazardous-materials suits.

The lines of authority and communication are established and clear, but everyone pitches in when required and sometimes when not. "We're a team," as Kyle said yesterday. They are; I've seen it. Extolling the virtues of the ship and her complement, I might sound like I'm trying to sell something. No, these are objective observations. I promise.

Videographer
Ben Harden,
not working.

• • • • • • • • • • • • • • • • •

We've just had a magnificent view of the east coast of Iceland, sets of high, shark-jaw mountains backlit by the sinking

sun, slashes of orange and soft pink reflecting on the bottoms of stacked cumulus clouds. And such sights as these—they're

another reason why people devote their nautical careers to research vessels.

LIFE'S ROUTINE, CONTINUED

Sun! It's actually shining; skies are blue, seas flat, nary a whitecap in sight, and the Weather Deck Secured signs (which mean "stay off") have been removed from the doors. The infernal rolling has largely ceased. The temperature soared to 10° C (50° F), wind 10 knots. We could hardly believe it. People were walking around the decks in T-shirts, sunbathing in Dan Torres's backyard hammock. This weather's supposed to hold for three whole days before the next inevitable low sweeps in.

We talked the other day about the bridge, the captain, and the mates' spheres of responsibility. Now let's go down one deck, via the "ladder," as stairs are called in nautical lingo, and work our way deck by deck to, finally, the engine room. The captain's cabin and the chief scientist's cabin share the 03 deck with the Shipboard Science Support Group (SSSG). Catie and Anton, heroes of the recent CTD incident, are responsible for everything electronic that collects science-related data—shipboard and lowered ADCPs, the multibeam sonar for mapping bottom topography, satellite communications, CTD sensors, computer servers. They serve as the bridge between the scientists and the permanent shipboard equipment while underway and before then, deciding what the oncoming scientists will need. That seems a lot for two people, but they pull it off, though for them this trip has been a bit more stressful than usual.

A leisurely Arctic afternoon.

On the 02 deck, one down, from forward to aft, we have the emergency-generator room, cabins port and starboard where crew and some of the science staff berth, the radio room, the library, and the so-called upper lab, now commandeered by those pesky outreach people. The radio room is home to Tony, the communications electronics technician ("com-ee-tee"). It's literally home; his cabin adjoins his office. You can always spot Tony; he wears shorts and T-shirts even out here north of the Arctic Circle.

Opposite: Viewing aft from the bridge at the industrial-strength cranes and communication pod.

Tony Skinner, communications electronics technician.

Kyle (*right*) in his bosun's locker.

He got his start at sea in 1980, when ships and shore were still communicating by Morse code and his digs would have been called the radio shack. Those days are gone, and now he maintains the bridge electronics, including the GPS systems, radars, gyrocompasses, fathometers, anemometers, and the complex dynamic positioning system. And if he needs to communicate with shore, any shore in the world, he picks up the satellite telephone.

And this brings us down another level to the main deck, which stretches the entire length of the ship and is, therefore, usually the one closest to the waterline. Up forward in its traditional position near the point of the bow, the "forepeak," is the bosun's locker. Kyle's domain is a storage room for about everything—paints, rope of all sizes, shackles, chain, and all manner of hardware. When not in use, the anchor chains, with hundred-pound links, are stowed below the bosun's locker, run up through huge, twin hawse pipes to the anchors mounted on either side of the bow.

Formally, the word is "boatswain," but no one says that now, if they ever did. Now it's bosun (the Skip calls him "swain"). The bosun is in charge of the deck and all activity that happens thereon. By the traditional cliché, the bosun, a jack-of-all-trades, is a big, tough, salty guy, not to be trifled with. When I first came aboard *Knorr* two trips ago, in Reykjavik, Kyle was the first man I met, the very model of a bosun, straight from central casting, flames painted on his hard hat. Well, he turned out to be a kind, gentle figure, the personification of the nautical work ethic, and still not to be trifled with, especially if you don't subscribe to that ethic. He oversees the OSs and ABs, who maintain the decks, houses, and all deck machinery, and who do indeed share his work ethic.

I think the central point of all this is the self-sufficiency that could only result from all these departments first doing their individual work and then working together.

Lunch in *Knorr*'s mess room.

The movie room, showings at 1900 every evening.

The upper lab.

A typical science-staff stateroom.

The movie lounge is aft of the bosun's locker—showings nightly at 1900. And that brings us to the most important spaces on the ship, the galley and mess room. Bobbie is chief steward, India is cook, and Tony is the messman. On this thirty-day cruise, Bobbie and India will prepare some 1,500 individual meals. For every lunch and dinner, they offer a meat or fish course with dessert and freshly baked breads, muffins, and coffee cakes, and special menus for vegetarians. Everybody's talking about the exceptional quality of the food. The galley staff puts in ten-hour days, but Bobbie and India take turns cooking lunch and dinner. Bobbie, who's been cooking at sea for over thirty years, the last twelve on research vessels, is also responsible for ordering all the stores before the trips and also for all the housekeeping items, everything from cleaning items to bedding, towels, pillows, napkins, and utensils.

Aft of the mess, the main deck is devoted to lab space, and aft of the main lab the stern is flat and open, the work platform for launching all manner of instruments, from moorings to bottom-coring equipment. The two decks below the main are called "platforms" for some reason. Most of the first platform is devoted to crew and science berthing. Carnival Line customers wouldn't be happy with the industrial style of the cabins (or the absence of alcohol), but they are perfectly adequate, comfortable, with bunk beds, plenty of stowage for personal gear, and adjoining heads and showers. There is a laundry room, a gym, and another lab space on the first platform. Below that, on the second platform, we find the muscle of the ship—the engine room.

Knorr has four diesel electric engines, three with sixteen cylinders, each producing 1,500 horsepower, and one eight-cylinder engine, "three cats and a kitten," as they like to say. To reduce fuel consumption, only two engines run at a time under usual working conditions. Unlike most ships, which transit from point to point at full diesel-power speed, *Knorr*'s missions don't require sustained high speed. The diesels don't directly drive the propellers—electrical motors do that, and the diesels produce just enough power to keep the electric motors running. Almost everything on the ship—ovens, water pumps, heaters, navigation and lab equipment, etc.—runs on electrical power, and because of the nature of *Knorr*'s work, priority goes to maintaining all these things, the so-called "house load," and secondarily to propulsion.

Engineers, then, are comparable to your local utility companies packaged into a convenient eight-person department. "Captain" of the department, Pete has three mates, called assistant engineers, first, second, third. (See "Engine Room Tour," on the enclosed CD.) They stand fixed watches, manning the engine room day and night, and, like the Skip's mates, they have particular responsibilities.

Chief Engineer Piotr (Pete) Marczak at his engine-room control panel.

One of *Knorr*'s three engines.

Russ Adams, the electrician.

The first assistant is in charge of the main engines, thrusters, pumps, hydraulic equipment, air-conditioning and refrigeration. Todd, the second assistant, is responsible for fuel and fueling (they burn the oldest fuel first) and the water makers (the desalinators can make 5,000 gallons of freshwater per day from seawater, but when in proximity to land, they stop making water for fear of loading polluted water.) Joe, the third, is in charge of the sanitation and sewerage systems, and he's responsible for the portable fire-fighting equipment. There are also three oilers, Nick, Ben, and Roger, who are comparable to the seamen working for Kyle. You see the oilers constantly making rounds checking this and that.

Last but not least, there is Russ, the electrician, a one-man department. He's associated with the engineering department, but mainly he's off on his own attending to anything electrical that doesn't fall into Tony's department or the SSSG's. Yesterday one of the ice lights, like headlights mounted on the bow tower, packed it in. Kyle, Jen, and Russ lowered the tower only to find that they lacked the necessary replacement part. This is no big deal for our trip, but next trip they'll be going to icy northern Greenland. So Russ managed to jury-rig something that will suffice, if needed, until they pick up the necessary part being shipped from Norway to Isafjördur. "Without Russ," said Pete, "it would be like missing an arm."

Most people never go down to the engine room, never even think of doing so. The engine room is sort of taken for granted. This, as Chiefski says, is a good thing. It means everything is working fine. However, it's also a good thing if, when we turn on our cabin lights, take a shower, or eat a hot meal, we recognize that it isn't the same as doing so on land and it isn't magic. It's because these guys are doing their job down there.

But I think the central point of all this is the self-sufficiency that could only result from all these departments first doing their individual work and then working together. That could be said about all seagoing ships. However, unlike most ships, *Knorr* goes to the ends of the Earth, to places where you can't visit the local marine-supply store to get what you need. You have to plan ahead, carry a lot of spares, and if something breaks, you've got to fix it yourself.

THE FAROE ISLANDS

Those two forces, fire and ice, along with a lot of erosion, have sculpted these islands.

Opposite: A small fishing vessel is dwarfed by the mountainside.

Far right, top and bottom: Sea stacks, remnants of once-great mountainsides, stand apart from the landmass.

The Faroes may be another of those places better revealed by imagery than language. So, unwilling to compete for fear of flamboyant failure, I'll offer but a few ash-dry sentences, eschewing the purple, and step aside to let the photographers evoke place, time, and light. That we're here in the Faroe Islands at all is a gift from Bob to the CTD team for their unstinting month-long labor.

The eighteen Faroe Islands, with a land area of 540 square miles and 694 miles of coastline, lie in the open North Atlantic midway between Iceland and northern Scotland at 62°00'N by 006°47'W, the position of the capital and largest town, Tórshavn. The islands are tightly clumped and separated by narrow fjords, suggesting that once there was a single island, perhaps of volcanic origin, arising, like Iceland, as a result of tectonic activity on the Greenland-Scotland Ridge. There's evidence in the rock of volcanism; and the passing of the Pleistocene glacier is etched into the landscape, evident not only in the fjords themselves but in other glacial features, cirques, couloirs, and arêtes. Those two forces, fire and ice, along with a lot of erosion, have sculpted these islands. But that's to get ahead of the story.

It began about 0800, landfall imminent, when the gawkers and the remarkers began to gather on the bridge, lining the bench at the after end, while Jen tried to navigate around us. She had chosen to round long, thin Kalsoy Island, lying at the north end of the group. The fog was thick, and the Faroes, still eight miles away, appeared first as a dark smudge, but even from that remove it became clear to the gallery that we were approaching a truly spectacular landscape. A half hour later, patches of blue sky struggled through the fog. Were we going to get lucky yet again, as on our approaches to Greenland and Iceland? Ten minutes later we variously gasped, whistled, or exclaimed profanely at an astounding piece of geology.

Top: Deep in the shadows a waterfall slides down this crevice.

Bottom: Six sheep traverse the steep mountainside.

A mountain—actually, half a mountain—soared up out of the black sea. It was a 700-meter-high vertical, naked-rock face, brooding wisps of off-white fog hanging near the summit, the timid sun glinting on the wet granite. On the landward side, the mountain eased down at a reasonable slope, but not on the seaward side. Something unimaginably violent had happened here in the deep past; something cataclysmic had cleaved away half of a mountain, leaving its innards exposed to the wind, the waves, and us. We fell silent; that doesn't happen often and never for long.

"That's where Grendel lives," said Pete after a time.

He was right. Geology, plate tectonics—no, myth, not science, best explains this place. A trio of thin, sharp crags, remarkably human-like, stood atop the cliff, as if from a Nordic pantheon in a souped-up Wagner set, brooding on the loss of half their aerie. At the foot of the cliff stood two huge sea stacks, cores from the lost half, like giant Rodin sculptures.

Yeah, but there were still serious nautical matters to attend to, and I was glad they didn't fall to me, leaving me and the gawkers to respond as we would to the spectacular, exotic land and seascape. The Skip had come onto the bridge with his inevitable mug of coffee. I noticed first what he didn't do. He didn't look over Jen's shoulder at the chart table, no second-guessing. Her navigation was good and he so assumed. (I love watching the way this ship works, if you haven't noticed.) He took his place on the starboard side. "What engines do we have on line?" he asked.

"One big and a small," said Jen.

"Let's put on another big." In close quarters, the loss of one big (sixteen-cylinder) engine, unlikely as it was, could put her on the ground.

We entered Kalsoyarfjødur, Kunoy Island on our port hand. (By the way, check some of these names, copied from the chart, of nearby places and features: Fuglafjørdur, Oyndarfjørdur, Funningsfjørdur, Gotuvk, and Haraldssund, evoking Vikings in longships and knarrs stopping for a rest en route to Greenland.)

In the fjord, the granite of Grendel's lair gave way to white sedimentary rock banding the green, pyramidal mountains like guitar strings. Then we began seeing farms and tiny settlements, each with a red-roofed church, nestled in the looping saddles between mountains. The shorelines on either side were steep-to, nowhere to land a boat. The isolation was exquisite.

"Jen, please give the engine room five minutes' notice for the bow thruster." *Knorr* is safe in open ocean. Danger comes when hard bottom is nearby. The Skip also asked Jen to call Kyle to ready the anchor for a quick drop. Kyle, whose birthday it was, spent the rest of morning standing by on the bow. Sightseers soon joined him. Others lined the rails; still others joined us on the bridge. A shoreline road appeared and periodically vanished into tunnels. How could they possibly have gotten construction equipment into these places, where the mountains sloped 45 degrees down from towering points? We saw barn-like buildings standing utterly alone in steep valleys seemingly accessible only by helicopter. I read subsequently that dried mutton was the national dish. Could these be drying sheds? But why were they built in such isolation, where the farmer risked life and limb to climb over the naked-rock mountain peaks on either side to reach his mutton?

From shore we must have looked like a cruise ship, photographers on the bridge wings, the flying bridge, and the main deck as we traveled down the fjord for over an hour past the mountains and the tucked-in towns of about twenty buildings each, all neat and spiffy. And then we stopped at the foot of the fjord between Kalsoy and much larger Søvágur Island in order to explore in the small boats.

Many of these waterfalls originate from within the mountains.

This page and opposite:
Such spectacular geography is
typical throughout the Faroes.

I was a member of the first boat tour, having lurked at the starboard side ready to go in life vest and hard hat as they craned the RHIB off of the 02 deck and placed it at the rail for boarding.

"Stay away from the town," said Kyle to Jose, the boat driver. There is an informal international agreement allowing research vessels to enter foreign waters for scientific work. Like Greenland, the Faroes are a self-governing dependency of Denmark, but we had no Danish clearance to land. So we sped off in the opposite direction from town to explore the wild, massive shoreline close aboard. There was a terrace of rock at the waterline where a boat might land, but landing would be pointless for the farmer trying to tend his sheep grazing farther up the steep mountainside, because he would be confronted by a sheer 50-meter cliff. Yet there were the sheep, grazing up on hair-raising slopes and seemingly impossible ascents to anywhere else.

Everywhere we looked, glimmering ribbons of waterfalls ran down ravines into the sea. The waterfalls emerged from out of the rock itself, not from the tops of the mountains, as

Blue-water grottoes undercut the rocks.

Joy riders under a rainbow. It rains 260 days a year.

The ocean will eventually claim these sea stacks.

though the islands had their own internal reservoirs—which is likely the case, since the Faroes receive 260 days of rain per year. Then, as we returned to the ship for the next riders' turn, a perfect rainbow of purple, orange, and yellow materialized from the clouds and arced over the mountain down to sea level as if Bob had staged it as part of his gift. And here was another of those pure, exquisite moments so valuable in experience yet so quickly passing into memory, precious and unforgettable.

That said, I'll step meekly aside and let the photographs portray the experience.

Everywhere we looked, glimmering ribbons of waterfalls ran down ravines into the sea.

"Grendel's lair."

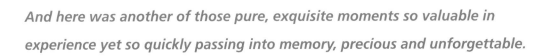

The Faroes are among the most isolated outposts in the North Atlantic.

A sea stack at the foot of Grendel's lair.

And here was another of those pure, exquisite moments so valuable in experience yet so quickly passing into memory, precious and unforgettable.

NEW CREWMAN, AII-EE!

Knorr's people make things look so easy, one can almost begin to think they are. But I know better, firsthand.

Knorr was parked bow into a moderate north wind and flat seas. The 255th CTD cast was complete. "Okay," said the Skip, "are you ready to put her on the next station?"

"Well…sure."

He told me to switch off the Dynamic Positioning (DP) inputs, take manual control and, using the joystick, come right to one-eight-zero. It's often compared to a video game, this DP system. I twisted the joystick, added some power, and watched the bow swing through 180 degrees. The Skip had walked me through the process last year in the Indian Ocean and earlier this trip, so I knew how to settle her on the new heading, engage the autopilot, and dial in fine-scale changes to her course; but I had never put her on station and held her there during an actual CTD cast. The next position was coming fast.

"So how are you going to make the turn?"

Yeah, I was wondering about that. I had to stop the ship on the waypoint with her bow pointing into the wind in the opposite direction. Skip suggested a looping turn to port about two ship lengths off the track to the waypoint, which appears as a blue line on the screen, then go past the waypoint and swing back into the wind. All this is visually displayed—a ship-shaped icon actually moving over the screen, leaving a "shadow" to show where she's been, and another icon indicating the place to stop.

"Start bleeding off speed…. Now turn her aggressively…. About a knot and a half now. Yeah, good." We were there. The Skip showed me how to engage the controls to keep her there. He informed

The author learning crane operation from Kevin.

Opposite: Heading back to work after our Faroes visit.

He pointed to the big radar screen. "Did you talk to this guy?" What? Oh. There was no target on the screen; we were alone. But I got the point. For all I had known, we were about to be T-boned by a supertanker.

the CTD watch and Kevin, the winch driver, that we were on station and "hove-to"—which is to say we were no longer moving through the water.

A common misconception about this fancy gear is that once you've told the ship to stay where she is—she will do so within a couple of meters—that's all there is to it. But we're putting the CTD package down perhaps 2,000 meters, where it can encounter currents different from those on the surface, causing the package to stream away from the hull, called "kiting," or, worse, inboard such that the wire abrades against the hull. So my task was to "keep the wire straight" by moving the ship in relation to the wire. I went out on the bridge wing to look at the wire as it went down. Good…. No, it was streaming aft.

"So what are you going to do?"

"Move her…aft."

"Let her settle in to be sure. You don't want to start chasing the wire." We watched…. "What do you think?"

"It's streaming aft. I'll give her a little reverse."

"Don't overdo it."

Just a tap of reverse on the joystick, and back out to watch the wire (don't chase it).

"Now let's talk about the scan. You don't want to focus on just one instrument. So what do you scan for?"

I ticked off the obvious factors: wind, current, position, wire. Remember, this only seems like a video game, the Skip had cautioned me. This is still a ship. Nautical forces are still going on out there.

He pointed to the big radar screen. "Did you talk to this guy?"

What? Oh. There was no target on the screen; we were alone. But I got the point. For all I had known, we were about to be T-boned by a supertanker.

So it went through the first cast: watch the wire, tweak the ship's position, scan the forces acting on the ship and the package until it's back aboard.

"Good job," said the Skip. I hope it's clear that, while I grasped the concept, I could not have executed the maneuvers without coaching.

Kevin, who'd been driving the winch, called on the "squawk box" to say that the CTD was secured.

"Which way are you going to turn for the next station?"

Almost ready to deploy…the author hopes.

"Starboard." Unlike the port side, the main deck on starboard is open. That's where all the work is done. Therefore, whenever possible, the bridge watch turns to starboard from the completed station toward the next, so as not to expose workers to the weather. I disengaged the DP controls and twisted the joystick to the right. We were underway again.

Kevin had come onto the bridge to assume the lookout, and so had Chief Mate Adam, whom the Skip was spelling during dinner.

"Dallas has got the con," said the Skip. That had a pleasing ring to it. If only it were true.

"Now, to get the whole CTD experience," said Adam, "you need to learn to operate the winch. And Kevin's the best teacher on the boat."

"Well…sure."

• • • • • • • • • • • • • • • • •

"Don't hit the bottom," said my teacher as we crowded into the winch-driver's hut on the 02 deck. "That's the first point. Don't two-block it. That's the second point." (I'll explain the two-block disaster in a minute.) He then explained the general process: You lower the package at 60 meters per minute—there's the speed indicator—until the operator on the control panel announces 100 meters from the bottom; then you slow the descent to 30 meters. At 10 meters from the bottom, you slow to a crawl until the operator says, "All stop." Then you start it back up at 60. At each stop, the operator records temperature, salinity, depth, and oxygen content, and they "fire" a bottle to capture a water sample.

A common misconception about this fancy gear is that once you've told the ship to stay where she is—she will do so within a couple of meters—that's all there is to it.

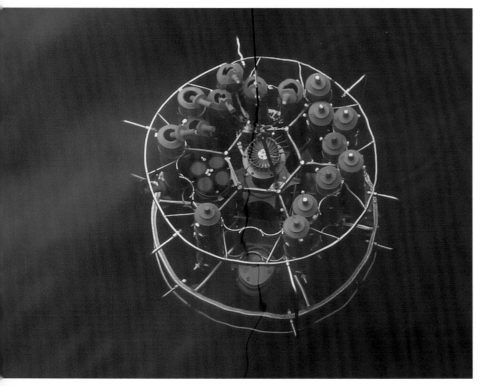

Still more CTD operations.

"You do that part." He didn't exactly say that any moron could do that part. "I'll launch and recover it, and you watch for a couple of casts, then you'll do it. Pay constant attention. That's the third point. You listen to headphones, read your Kindle—that's when accidents happen and people get hurt."

I watched Kevin launch and retrieve the CTD package—the hard parts—while I ran the easy middle part. We paid close attention to the wire-out meter and winch-speed meter while we discussed literature we liked. "Okay," he said, "are you ready to run the next one?"

"Ready." If not able.

Here's the hard part: The winch lifts the CTD straight up off the deck, the wire running though a block, or pulley, around a couple of fairleads to the big winch drum next to our hut. However, when it goes straight up, it's still over the deck, not the water. To get it over the water, the winch driver has to extend the heavy boom to which the block is shackled. Trouble is, extending the boom also acts to take in wire, thus lifting the CTD toward the block. That's where the two-block disaster looms. If the driver doesn't compensate by lowering the package as he extends the boom, then the block at the top of the package frame will be jammed up into the big winch block—two-blocked—and this could break the wire. And a couple of hundred thousand bucks' worth of kit goes to the bottom. "Got it?"

"Uh…so as you extend the boom, you've got to lower the package at the same time. Both at once."

"Right. And you do the opposite as you retrieve it. You do it next time. I'll be right here." To pick up the pieces.

My time unavoidably arrived. I stood around the winch hut with Kev trying to look nonchalant as we came onto station and stopped. Then I took my place in the hut looking down at the CTD and Kate and Stine, who were unstrapping it from the pallet. Good luck, they said. That's when I noticed an audience was gathering. Catie stood on the 02 deck. Was she smirking? The Skip came on deck to watch. Others materialized on the main deck.

Could I get out of this? When the boom goes out, the package has to go up. No, no, it's the other way around. Concentrate. The package has to go down when the boom goes out. Down…not up. Up on the retrieval, when the boom comes in. Sorry, Kev, gotta go. Jury duty. Forgot to mention that.

I called the bridge for permission to launch, with scant hope that they'd deny it. Up when the… I put the winch lever forward, thereby unspooling great snakes of wire onto the deck.

"Stop!" Kevin squealed.

Anton and Catie held the wire taut while Kev respooled it. I didn't turn around, picturing the audience seeking shelter indoors and/or summoning the Damage Control team.

"Try it again, only this time lift it first."

I took both levers in sweaty hands, hoisted the package off the deck above the rail height. This was no time to quit smoking. Now, the boom out, the package down to compensate—well, to make an anxious story short, I didn't make a further hash of it. I got it down and back up without incident or injury, but not much beauty, and thank God for the long steam to the next station. I needed a shower.

"So what job are you going to do next?" asked some wag as I walked into the mess deck.

"Captain?"

The author has (haltingly) lowered the package.

HYPOTHESIS: CONFIRMED

Scientists tend toward professional, if not temperamental, prudence, skepticism. There's a running joke between Bob and me about our very pleasing collaboration. When I finish one of these essays that relate to science, I pass it to him for approval. He qualifies nearly every categorical statement, adding "usually," "typically," "mostly." But there is a fundamental point here. In science, the highest form of certainty is the theory; after all, new data may come to light or old data may be reinterpreted, a point fundamentalists and ideologues have exploited to say that evolution, for instance, is "just a theory." However, Bob and Kjetil are prepared now to claim unequivocally that they have confirmed their hypothesis. Before we get to how and why, it's probably worthwhile to restate the hypothesis.

Most ocean currents flow from somewhere toward somewhere else. They have no beginning or ending, except insofar as humans have named them for their own ease of reference, but are parts of a continuous circulation. However, the North Icelandic Jet is different. It does not flow from elsewhere over the top of Iceland thence into the Denmark Strait. Instead, it is formed right here in the waters on the north slope of Iceland. Here's how, per the hypothesis:

A branch of the warm North Atlantic Current, the North Icelandic Irminger Current (IC), flows northward, transporting some one million cubic meters of warm water per second northward through the Denmark Strait, and then bends to the east over the top of Iceland. There the IC dies out, devolving into eddies that spin energetically out into the Iceland Sea. Relinquishing its heat to the cold Arctic atmosphere, the once-warm water grows dense and slips beneath the surface, joining a large reservoir of cold water in the Iceland Sea. So we have a quantity of water moving north; and now we need to think circles. Nature loves circles because circles conserve mass. If a quantity of water moves north, then an equal quantity must move south. The laws of physics so require. As water from

In the ocean may lie the answers to questions humankind has not yet learned to ask.

Opposite: The fog lifts near Cape Tupinier, Greenland.

the cold, dense reservoir "replaces" the Irminger Current, it sinks—becoming the North Icelandic Jet—and then flows south into the Denmark Strait. That's the hypothesis this cruise set out to test.

To verify the hypothesis, Bob and Kjetil needed to demonstrate two things. One, that the NIJ water did not flow from somewhere else. And two, that if the current was formed right there on the northeast slope of Iceland and then flowed southwestward into the Denmark Strait, it had to "disappear" at some point upstream.

As to the first point, please see Figure One, the map of Iceland. Note CTD sections 4, 15, and 16, which cover other likely places the current could be flowing from. In these, Bob and Kjetil found no traces of the NIJ. One down.

Now, as to the second point, please see Figure Two. These eight panels, which are generated from CTD and ADCP data, depict current velocity. The "line numbers" 2 through 10 refer to the CTD sections on Figure One. The yellow color represents the NIJ itself. Notice that, beginning in the Denmark Strait, lines 2 and 3, the current is zipping right along. After line 3, it lessens in velocity, and it continues to diminish through line 8. By lines 9 and 10, on the northeast of Iceland, it is gone.

The evidence could hardly be clearer: Hypothesis confirmed. However, to produce this end result required several voyages and countless hours of combined intellectual effort by four physical oceanographers: the Icelandic scientists who noticed the "mysterious current" in the first place and Bob and

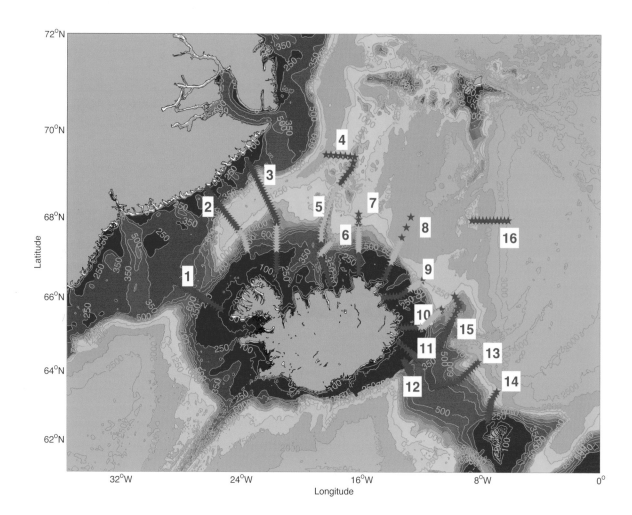

Figure One

This shows Bob's entire study area. Cyan stars denote the sections shown in Figure Two.

To verify the hypothesis, Bob and Kjetil needed to demonstrate two things. One, that the NIJ water did not flow from somewhere else. And two, that if the current were formed right there on the northeast slope of Iceland and then flowed southwestward into the Denmark Strait, it had to "disappear" at some point upstream.

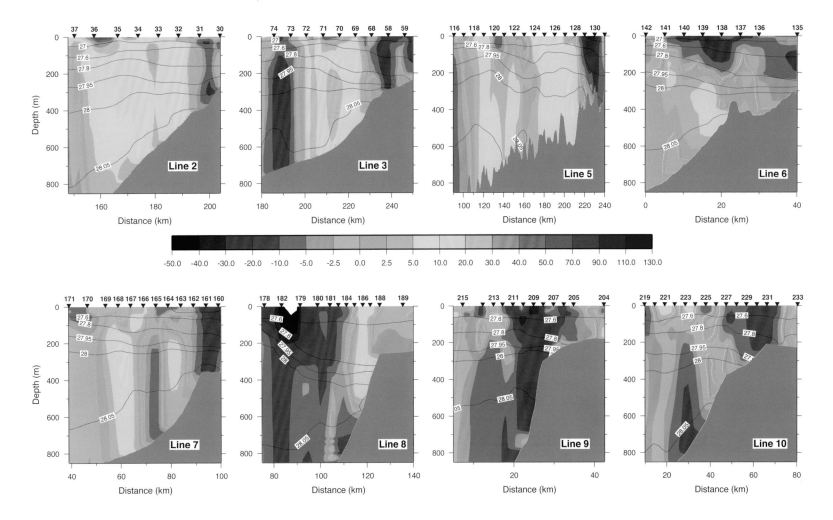

who confirmed its existence in 2008 and now have found its origin, plus the modeling work by Michael Spall. Further, if oceanographic data were water itself, Bob and Kjetil would be neck deep in it. They have loose-leaf binders full of velocity data from lowered ADCPs and shipboard ADCPs plus that from 335 separate CTD casts. Further, the discovery of the North Icelandic Jet and now of its origin introduces a new set of questions—for example, what causes the NIJ to flow into the Denmark Strait instead of the other direction. To answer that and other questions will require more fieldwork and modeling.

So what does it mean, all this to-do about a relatively small current in a remote part of the foggy, stormy Nordic Seas north of the Arctic Circle where almost no one ever goes? The oceanographer's job, like that of most other Earth scientists, is to learn and to explain how nature works. That's it. But that's a lot.

Figure Two

These eight panels, which were generated from CTD and ADCP data, depict current velocity. The "line numbers" 2 through 10 refer to the cyan CTD sections on Figure One. The yellow/orange colors represents the NIJ itself. Notice that, beginning in the Denmark Strait, lines 2 and 3, the current is zipping right along. After line 3, it lessens in velocity, and it continues to diminish through line 8. By lines 9 and 10, on the northeast of Iceland, it is gone.

The evidence could hardly be clearer: Hypothesis confirmed.

That's what Galileo, Newton, and Darwin did, but such "big" discoveries have already been made, at least theoretically. More modest questions about how the world works remain unanswered, and in the ocean many remain.

Further, the dynamics of these northern seas are particularly important in light of climate change. The North Icelandic Jet stands as a new mechanism by which dense water is transported southward to complete the north-south circulation on which the stability of our climate depends. It delivers fully half the water flowing southward through the Denmark Strait to form the headwaters of that return flow. It's certainly not out of the question that, by a system of yet-unknown feedbacks, anthropogenically caused climate change could modify, perhaps even sever that circulation. We can't possibly understand how and if that might happen without understanding the mechanisms within that oceanic circulation. And in the ocean may lie the answers to questions humankind has not yet learned to ask.

Knorr moves off the Blosseville Coast of Greenland with a fulmar as escort.

An old iceberg grounded off the East Greenland coast.
It might have been adrift for decades.

AN ENDING

The sky sulks, there's a light drizzle, and a long, lazy swell is rolling around from nowhere in particular. It's a melancholy day, our second-to-last, and the mood is again either enhanced by or reflected in the weather. I'm sorry to see this cruise end. I'm not alone in that. Even the Captain and members of the crew, some of whom have been aboard for four months, expressed similar sentiments this morning. None of us will ever forget the spectacular violence during the 2008 cruise, 45-foot waves with breakers boiling on their crests, but for the Arctic sights we've seen, for the unusual nature and high level of scientific success, this cruise will be hard to beat.

We stopped twice in different north-Iceland fjords. We came close aboard the unforgettable Blosseville Coast of East Greenland and later the improbably exotic Faroe Islands. In each place we deployed small boats in the service of documentary photography. None of *Knorr's* old guard remembers so many small-boat deployments, and some said they dreaded the near approach to land for fear the outreach group would want to do it again. During the first part of the cruise, we deployed twelve moorings in the Denmark Strait, where they will remain for a full year collecting temperature, salinity, and velocity data about this vital arc in the North Atlantic's Meridional Overturning Circulation. When we return next August to recover the moorings and retrieve their data, we'll be a giant step closer to understanding the dynamics of the ocean in the eastern Arctic. It is not to diminish its importance or the skill required to execute it to say that mooring work is typical of most physical-oceanography expeditions. The second part of the cruise was far from typical.

Opposite: Ísafjördur, Iceland, where it ended.

The pursuit of ocean knowledge was downright literary. It had structure—a clear objective, a search, and a climactic conclusion. It obeyed the unity of time; when the cruise was complete, so was the story. Even now I can hear Bob saying he doesn't want to leave the impression that the work is done.

A final night's treat—the aurora borealis.

It's far from done. Bob and Kjetil are awash in data yet to be understood. Their discoveries have ushered in a new set of questions about just what goes on beneath the surface in this dizzyingly complex region; that story will likely be told in specialized jounals. But our exploration story is complete, and in miniature it is the story of the advance of science.

The Icelanders caught a snapshot signal of something moving where movement was never noticed before, and they were shrewd enough to take it for what it turned out to be, a hitherto unknown current. With careful, high-resolution measurements aboard this same ship, Bob and Kjetil subsequently verified the existence of the North Icelandic Jet, but that left hanging the question, where did it come from? Or if it didn't come from anywhere, how did it come to exist?

Thirty days after leaving Reykjavik, after covering 3,812 nautical miles, collecting real-time data from 335 CTD casts and reams of velocity data from the ADCPs, after spending some 14 hours a day in headphone isolation at their computers, Bob and Kjetil found the origin of the NIJ. It was right here, just as they'd hypothesized, in the waters on the north slope of Iceland, and it delivered half of the total quantity of water flowing southward through the Denmark Strait. Like all good quest stories, this one ends decisively. That's unusual in at-sea oceanography.

Ísafjördur, Iceland.

The harbor pilot's arrival traditionally marks the end of a voyage.

The Icelandic pilot confers
with Captain Sheasley.

Unlike many popular sea stories, ours contained no injuries, no loss of life to the heartless ocean

(one guy lost a filling), no privation, and no cannibalism. And now the end is literally in sight. Most

of both watches are up and about at 0730. We're supposed to pick up the harbor pilot in an hour, but

the big cruise ship presently occupying his attention could delay that. The Skip's on the bridge with

Jose, the helmsman, Second Mate Jen, and the inevitable spectators. Kyle and the deckies have laid

out the mooring lines, tied on the heaving lines, readied the anchor and the gangway. Now Kyle's in

the bosun's place on the bow with Chief Mate Adam; Mike, the third mate, is on the stern to oversee

line handling. These people who approach things with relaxed, joking informality are now, when it's

necessary, straight-up, serious professionals at work. It's pleasing to watch, but sad, too. We're leaving

the ship in Isafjördur, and I don't know when I'll be back.

Docking, the actual end of a voyage.

The Skip and his crew neatly placed her starboard side against the industrial wharf lined with metal-sided warehouses. ("These are the same harbor-side buildings I see wherever I go," Kevin said, "I think they just move them from place to place.") The heaving lines arced onto the concrete wharf; the heavy hawsers followed; the gangway was craned into place and made fast; the pilot left the ship. That's as over as a cruise gets.

Still, I'm standing around on the side deck talking with the crew to prolong the leaving. Most of the science staff have already left for the airport. The mates and the engineers are going through the changeover procedures, and I'm drinking another cup of coffee. But I know that all sea stories must close. *Knorr's* people have treated me like one of their own, a privilege I'll not forget. And after copping in one of the first essays to the tendency toward sentimentality about the ocean and ships, particularly *Knorr*, I promised to repress it. The thought of not seeing her again is too sad to brook without breaking the promise. Bon voyage to *Knorr* and all who sail in her.

Celebrating Knorr ashore.

SCIENCE TEAM

Robert Pickart
Chief Scientist
Woods Hole Oceanographic Institution,
Massachusetts, USA

Hédinn Valdimarsson
Co-Principal Investigator
Marine Research Institute, Reykjavik, ICELAND

Kjetil Våge
Co-Principal Investigator
University of Bergen, NORWAY

Laura de Steur
Co-Principal Investigator
Royal Netherlands Institute for Sea Research,
Texel, NETHERLANDS

John Kemp
Mooring Technician
Woods Hole Oceanographic Institution,
Massachusetts, USA

Jim Ryder
Mooring Technician
Woods Hole Oceanographic Institution,
Massachusetts, USA

Dan Torres
Acoustic Doppler Current Profiler Technician
Woods Hole Oceanographic Institution,
Massachusetts, USA

Carolina Nobre
CTD Processor
Woods Hole Oceanographic Institution,
Massachusetts, USA

Dave Wellwood
Hydrographer
Woods Hole Oceanographic Institution,
Massachusetts, USA

Lena Schulze
Conductivity-Temperature-Depth (CTD) Team
Woods Hole Oceanographic Institution,
Massachusetts, USA

Stefanie Zamorski
CTD team
University of Rhode Island, Graduate School
of Oceanography, Rhode Island, USA

Kate Lewis
CTD team
Woods Hole Oceanographic Institution,
Massachusetts, USA

Ashley Stinson
CTD Team
University of Maine, Maine, USA

Stine Hermansen
CTD team
University of Bergen, Bergen, NORWAY

Mirjam Glessmer
CTD team
University of Bergen, Bergen, NORWAY

Leon Wuis
Mooring Technician
Royal Netherlands Institute for Sea Research,
Texel, NETHERLANDS

Magnus Danielsen
Mooring Technician
Marine Research Institute, Reykjavik,
ICELAND

ACKNOWLEDGMENTS

I would to like express my deep appreciation to Dr. Eric Itsweire, head of the Physical Oceanographic Section of the National Science Foundation (NSF) Ocean Sciences Division, for his enthusiastic support of outreach. Without his encouragement and the financial backing of NSF, we would not have been able to document the cruise and produce this book. I am lucky and thankful that Eric believes so strongly in the merit of communicating our exciting results to the general public.

I am thankful to the Ocean and Climate Change Institute at the Woods Hole Oceanographic Institution (WHOI), and in particular Dr. William Curry, for his support of high-latitude research. Bill has continually helped me "move north" during my career and has provided opportunities that ultimately helped shape this field program in the Iceland Sea. Jim and Ruth Clark established the "Arctic Research Initiative" at WHOI, and through their enormous generosity enabled me and many others to explore new ideas in the northern climes. WHOI's development office, in particular Jane Neumann, Priya McCue, and Annamarie Behring, have been extraordinarily enthusiastic and supportive of my research and outreach. Gratia "Topsy" Montgomery shared her love of the ocean with me and helped me enter the world of the Arctic. Dick Pittenger, former head of WHOI Marine Operations, has a true passion for adventure at sea and has been instrumental in allowing me to experience such adventure. I would also like to thank former WHOI Director Dr. Robert Gagosian and former WHOI Director of Research Dr. James Luyten, as well as current WHOI Director/President Dr. Susan Avery and current Director of Research Dr. Larry Madin.

As always, it was a pleasure to sail on R/V *Knorr*. Not only did Captain Sheasley and his crew carry out our science operations in a safe and effective manner, but they bent over backwards to accommodate our outreach activities—always willing to launch the small boat at a moment's notice, answer questions in front of a camera, write emails to schoolchildren. They clearly love what they do, and, thankfully for us, were happy to share their thoughts and experiences with us.

Lastly, my wife Anne has made this all possible: "holding down the fort" on my long trips to sea and enabling me to explore the far corners of our oceans while always having a loving family to return home to.

Robert Pickart, Chief Scientist

PHOTO CREDITS

Front Cover: Rachel Fletcher

Back Cover: Rachel Fletcher

Opposite Copyright and Contents: Rachel Fletcher

Introduction: *Opposite p. i:* Earthstar Geographics LLC;
pp. i–ii: Sindre Skrede; *p. iii:* Rachel Fletcher

Journal 1 *Opposite p.1, p.2:* Rachel Fletcher; *p.4:* Sindre Skrede

Journal 2 *pp. 4, 6, 8, 9:* Rachel Fletcher

Journal 3 *pp. 10–15, 16:* Rachel Fletcher

Journal 4 *pp. 18, 21:* Rachel Fletcher

Journal 5 *pp. 22, 23:* Rachel Fletcher

Journal 6 *pp. 26–29:* Rachel Fletcher

Journal 7 *pp. 31–36:* Rachel Fletcher; *p. 37 bottom left:* Sindre Skrede;
far right, top and bottom: Rachel Fletcher

Journal 8 *pp. 38, 40–45:* Rachel Fletcher

Journal 9 *pp. 46–49:* Rachel Fletcher

Journal 10 *pp. 50–51:* Sindre Skrede; *p. 53:* Rachel Fletcher

Journal 11 *pp. 54, 56:* Rachel Fletcher; *p. 57:* Sindre Skrede

Journal 12 *pp. 58–60, 62–63:* Sindre Skrede

Journal 13 *p. 64:* Rachel Fletcher; *pp. 65, 68–69:* Sindre Skrede

Journal 14 *p. 70:* Rachel Fletcher; *p. 71 top and bottom:* Sindre Skrede;
p. 71 top and middle: Sindre Skrede; *bottom:* Rachel Fletcher; *p. 73:* Sindre Skrede

Journal 15 *pp. 74–75, 79:* Sindre Skrede

Journal 16 *p. 80:* Rachel Fletcher; *p. 81, top:* Sindre Skrede; *bottom:* Wayne Silva
p. 82, top: Sindre Skrede, bottom: Rachel Fletcher; *p. 83:* Sindre Skrede

Journal 17 *p. 86 top left:* Rachel Fletcher; *p. 88, top right:* Sindre Skrede;
bottom: Rachel Fletcher; *p. 89, top:* Sindre Skrede; *bottom:* Rachel Fletcher;
p. 90 top left, below left, bottom: Rachel Fletcher; *top right:* Sindre Skrede;
p. 91: Rachel Fletcher

Journal 18 *pp. 92–93:* Rachel Fletcher; *p. 94, top:* Sindre Skrede;
bottom: Rachel Fletcher; *p. 95, top right and left:* Rachel Fletcher;
middle and bottom right: Sindre Skrede

Journal 19 *pp. 98–107:* Rachel Fletcher

Journal 20 *p. 108:* Sindre Skrede; *pp. 109–111:* Rachel Fletcher

Journal 21 *p. 114:* Rachel Fletcher; *p. 118:* Rachel Fletcher;
p. 119: Sindre Skrede

Journal 22 *p. 120:* Rachel Fletcher; *pp. 121–122:* Sindre Skrede;
p. 123: Rachel Fletcher; *p. 124 top:* Sindre Skrede; *bottom:* Rachel Fletcher;
p. 125: Sindre Skrede; *pp. 126–127:* Rachel Fletcher

Page 132: Rachel Fletcher

INDEX

References in boldface indicate photographs.

And so good-bye and farewell—for now—to the Arctic.